NEW TRENDS IN BIOCHEMICAL PHYSICS RESEARCH

NEW TRENDS IN BIOCHEMICAL PHYSICS RESEARCH

SERGEI D. VARFOLOMEEV
ELENA B. BURLAKOVA
ANATOLII A. POPOV
AND
GENNADY E. ZAIKOV
EDITORS

Nova Science Publishers, Inc.
New York

For permission to use material from this book please contact us:
Telephone 631-231-7269; Fax 631-231-8175
Web Site: http://www.novapublishers.com

LIBRARY OF CONGRESS CATALOGING-IN-PUBLICATION DATA
Available upon request

ISBN 13 978-1-60021-463-9
ISBN 10 1-60021-463-0

Published by Nova Science Publishers, Inc. ✣ New York

CONTENTS

PREFACE

This book presents recent research in chemical and biochemical physics. Chemical physics addresses a large range of problems. An effective chemical physicist is a "jack-of-all-trades," able to apply the principles and techniques of the field to everything from high-tech materials to biology. Just as the fields of chemistry and physics have expanded, so have chemical physics subject areas, which include polymers, materials, surfaces/interfaces, and biological macromolecules, along with the traditional small molecule and condensed phase systems. Biochemical physics is a science that joins the three natural sciences (biology, chemistry and physics) into one comprehensive study. N.M. Emanuel pioneered this science over fifty years ago. This book brings together papers written by Emanuel's students, which reveal recent developments in this interesting field.

Chapter 1 - The morphology and the thermal and mechanical properties of ternary blends prepared by melt mixing low density polyethylene (LDPE), isotactic polypropylene (iPP) and a ethylene-propylene copolymer (EP) were investigated. Differential scanning calorimetry (DSC), small-angle (SAXS) and wide-angle (WAXS) X-ray scattering, scanning electron microscopy (SEM) and large tensile deformation tests were performed for blends containing three different weight fractions of EP (5, 10 and 15 wt.-%) and the results are compared with the behaviour of the binary LDPE/iPP blend.

Chapter 2 - The historical dynamics of men's and women's lifespan differences is analyzed. Regular changes of men's overmortality during different historical periods are found. In a number of developed countries significant growth of men's overmortality with subsequent decrease during the period 1960 through 2000 is found. The start and the end of the men's overmortality peak differ in different countries. In some countries the decrease of the men's overmortality has not begun yet, but the preceding dynamics coincides with the dynamics in the other countries, which allows predicting the changes of the men's mortality in the near future.

Chapter 3 - At contact of ozone with polypropylene fiber (PP) ozone quickly enough reacts with it. Reaction proceeds in surface nanolayer (10^{-9}м), thus many properties of a fiber essentially change. The complex of experimental works in which changes of physical - chemical properties of PP fiber and nonwoven term bound polypropylene material received during its reaction with ozone-oxygen mix are investigated is executed. It is established, that reaction proceeds in surface layer of a fiber and is accompanied by substantial growth of the PP fibers module of elasticity at break (23 %), and also significant, increase in a moisture capacity.

Chapter 4 - Thermodynamic incompatibility is a fundamental physico-chemical phenomenon which determines the structure, rheology and functional properties of (bio)polymer mixtures [1]. Biopolymer mixtures are of importance to control the composition –structure-property relationship in natural biological systems and in formulated food. Therefore, to predict the properties of mixed biopolymer systems, their phase state and phase diagram should be known over a wide range of physico-chemical parameters [2].

The aim of the work is an investigation of the phase state and the phase behavior of the ternary water-methylcellulose (MC)-locust bean gum (LBG) system, determination and description of the phase diagram using the main thermodynamic parameters obtained.

Chapter 5 - A detailed quantum-mechanical analysis of a model of a water-oxidizing complex, based on recent X-ray data on the photosystem II structure, is presented. For the first time, a mechanism of water oxidation is proposed and explained within the framework of this model. The role of three manganese atoms, which are not directly involved in water oxidation; the role of cubic structure of the complex; and the necessity of Ca and Cl ions for water oxidation are discussed. Theoretical computations of energies of the complex were made for each of the *S* states.

Chapter 6 - This paper summarizes the introduction of low dielectric constant (low-k) materials in microelectronic devices and points out main application problems.

After a broad overview of new low-k materials the authors focus on introduction of porosity as a key approach to lower down permittivity. Chemical methods to produce porosity by incorporation of porogens, use of solid state solvents and thermal degradation with carbon radicals are discussed.

The main challenge lies in generating low-k dielectrics compatible with modern semiconductor processing. The authors study two versatile physical methods to characterize mechanical and structural properties: nanoindentation and laser-generated surface acoustic waves.

Chapter 7 - Diversion of organic waste from land fill or incineration, to biological treatment by anaerobic digestion is of current interest in Austria. Now energy production from sewage sludge, solid wastes and energy crops by anaerobic digestion is wide spread practiced. The objective of this study was to optimise anaerobic treatment of different kinds of organic waste in a technical scale digester. New methods and laboratory analysis, for the description of microbiological process, were investigated and used. Detailed data have been collected from approximately 150 plants, which generate green electricity by means of energy crops, manure and organic waste. Based on the collected data from the biogas plants, a benchmark system for required process parameters was developed. This system helps the plant owners to operate their plants under monitored conditions in an economical way.

In: New Trends in Biochemical Physics Research
Editors: S. D. Varfolomeev et al., pp. 1-7
ISBN 1-60021-463-0

Chapter 1

MORPHOLOGY, THERMAL AND MECHANICAL PROPERTIES OF TERNARY LDPE/PP/EP BLENDS

***O. Priadilova*[1,3], *N. M. Livanova*[3], *O. A. Ledneva*[3], *A. A. Popov*[3] *and J. S. Gutmann*[1,2]**

[1]Max-Planck-Institute for Polymer Research, Mainz, Germany
[2]Institute of Physical Chemistry, University of Mainz, Mainz, Germany
[3]Institute of Biochemical Physics, Russian Academy of Sciences, Moscow, Russia

ABSTRACT

The morphology and the thermal and mechanical properties of ternary blends prepared by melt mixing low density polyethylene (LDPE), isotactic polypropylene (iPP) and a ethylene-propylene copolymer (EP) were investigated. Differential scanning calorimetry (DSC), small-angle (SAXS) and wide-angle (WAXS) X-ray scattering, scanning electron microscopy (SEM) and large tensile deformation tests were performed for blends containing three different weight fractions of EP (5, 10 and 15 wt.-%) and the results are compared with the behaviour of the binary LDPE/iPP blend.

Keywords: blend, polyolefins, mechanical properties, morphology

INTRODUCTION

Polyolefins remain the most important class of commodity plastics and polyolefin-based blends have been widely studied polymer systems from scientific as well as industrial interests. [1] There have been many discussions about compatibility in polyethylene/ polypropylene blends and related mechanical properties with respect to the blend composition and the microstructure of both homopolymers. [2] It is well-established that polymer/polymer miscibility is determined by a delicate balance of enthalpic and entropic forces. However, for high molecular-weight materials, the combinatorial entropy term that explains the miscibility observed in solvent mixtures and polymer solutions is very low and polymer/polymer miscibility is primarily controlled by the nature of the interactions between the chemically different macromolecules. In most of cases, polymer pairs are immiscible giving phase-separated morphologies. From an experimental approach, the immiscible character of polymer blends can be easily evidenced by investigating for instance, the morphology of the blends by means of electron microscopy technique. This microscopy method has been largely extended to the study of compatibilization of immiscible binary blends upon addition of third component leading to a new ternary blend usually referred as compatibilized blend. [3] Usually such compatibilizers have blocks or graft segments that are chemically identical to those in the respective phases and work by improving the interfacial adhesion. This strategy was inspired from colloid science where the idea of adding a molecule having hydrophobic and hydrophilic moieties, i.e. surfactant, would allow dispersing and stabilizing emulsions of water-in-oil or oil-in-water. In the present paper, we compare the morphologies, thermal and mechanical properties of melt-blended low density polyethylene/isotactic polypropylene (LDPE/iPP-80/20) blends modified by the addition of an ethylene-propylene copolymer. The experimental results are discussed with respect to the ethylene-propylene copolymer content.

EXPERIMENTAL

Materials and blend preparation Low density polyethylene (brand Kaplen 158-03 020) and isotactic polypropylene (brand Kaplen 01025) were purchased from MPZ company (Moscow, Russia). Ethylene-propylene copolymer with ethylene content of 64 mol-% and M_w=200 000 g.mol^{-1} was provided by DSM (Netherlands). Blends with a fixed LDPE/iPP weight ratio (80/20) and varied EP weight content (5, 10 and 15 wt.-%) were prepared by melt-mixing at 190°C for 10 min with a Brabender mill. The ternary blends are referred as LDPE/iPP/EP-*x* where *x* refers to the EP content. The resulting blends were compression moulded at a temperature of 190°C to produce films with well controlled thickness.

Differential scanning calorimetry DSC measurements were performed on a Mettler DSC 30 apparatus with a cell purged with nitrogen at a temperature variation rate of 10°C.min^{-1} for both heating and cooling runs.

Small-angle and wide-angle X-ray scattering WAXS measurements have been preformed using a θ-θ diffractometer (*Siemens*). Cu Kα radiation (λ=0.154 nm) was used. For the SAXS measurements a rotating anode X-ray source with with a pine-hole beam collimation and a 2D position sensitive detector were used. The measurements have been preformed at room temperature. The recorded scattering intensity are reported as a function of the scattering vector ($s=2\sin\theta/\lambda$, where 2θ is the scattering angle).

Scanning electron microscopy SEM images were obtained using a LEO EM 1530 Gemini scanning electron microscope. In order to observe the bulk features of the blends, compression moulded films having a thickness of about 400 μm were fracture in liquid nitrogen and the fractured surfaces were examined.

Large deformation tensile tests Measurements were performed using a mechanical testing machine (INSTRON, Model 6022). Sample specimens having a thickness of about 80 μm were cut with a dumbbell shape in which the gauge length was 25 mm. The stress-strain curves were measured at a constant cross-head speed of 10 mm.min^{-1}. Young modulus (E)

and the coordinates of the yield point as well as of the elongation at break have been determined as averages of at least 7 independents drawing experiments performed in the same conditions. The meaning of the parameters measured is illustrated in Figure 1. Only the room temperature behaviour is reported.

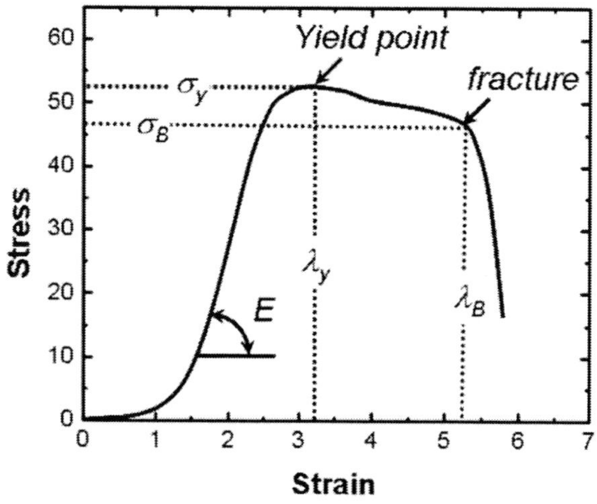

Figure 1. Schematic stress-strain dependence typical for a semi-crystalline thermoplastic. The parameters characteristic of the elasticity at small deformation as well as the yield and the break points are indicated.

RESULTS AND DISCUSSIONS

According to the producer, the ethylene-propylene copolymer used in this study is an ethylene rich copolymer with a random distribution of the ethylene and propylene units along the polymer backbone. The rubbery behaviour of the ethylene-propylene copolymer is expressed by the changes in heat flow and mechanical loss factor (tan $\delta = G''/G'$ where G' is the storage moduli and G'' is the loss modulus) versus temperature presented in Figures 2a and b, respectively. A good agreement is obtained for the T_g values determined by both thermal and mechanical methods. Blends of LDPE and iPP are known to behave as immiscible systems with variation depending on the microstructure of the polymeric components. This phase segregation behaviour can be recognized from the DSC data of Figure 3a and 3b presenting the thermograms recorded for binary LDPE/iPP and ternary LDPE/iPP/EP blends during cooling from the melt and subsequent heating processes, respectively. The four blends exhibit two melting peaks and the temperatures correspond well to the melting transition peaks recorded for the homopolymers. This latter observation suggests that the interactions between the crystalline domains of different chemical nature are very limited. It should also be noticed that the crystallization of the iPP component is greatly affected when embedded in a LDPE matrix. In the bulk semi-crystalline polymers, nucleation process usually occurs through heterogeneous nucleation, i.e. nucleation on existing heterogeneities such as impurities, catalyst debris... There have been many debates about the delayed crystallization of dispersed phase domains in immiscible blends and in most cases this phenomena has been related to the lack of sufficiently active heterogeneities within the phase segregated regions. [4] The spherulitic morphology of melt-crystallized LDPE/iPP and LDPE/iPP/EP-*x* samples was evidenced by polarized optical microscopy and 2D small-angle light scattering. Volume filling spherulitic morphology and anisotropic light scattering pattern were obtained in all cases suggesting that the amorphous ethylene-propylene copolymer is located within the semi-crystalline superstructures. The semi-crystalline/semi-crystalline morphology of LDPE/iPP and LDPE/iPP/EP-*x* blends was further investigated by X-ray scattering experiments performed in both the wide-angle and small-angle regimes. The combined X-ray results presented in Figure 4 are consistent with the semi-crystalline nature expected for LDPE/iPP based blends investigated in this study. The peaks in the SAXS profiles result from the density contrast between amorphous and crystalline domains. The WAXS intensity distributions showing diffraction peaks superimposed to a diffuse halo are also characteristic of this biphasic character. More detailed examination of our overall WAXS data indicated that crystalline forms typical for LDPE and iPP homopolymers were also detected for both binary and ternary blends (representative examples are given for LDPE/iPP and LDPE/iPP/EP-15 in the inset of Figure 4). Together with the data presented above, the shift of the peak position in the SAXS profile suggests that the EP copolymer partly is incorporated within the crystalline domains of the blends. A first inspection of SEM micrographs shown in Figure 5 reveals in all cases a two-phase morphology consisting of iPP dispersed domains with a nearly spherical shape dispersed in a LDPE matrix. A more careful inspection reveals a dramatic reduction of polypropylene domains size upon EP addition. Quantitative analysis of SEM pictures revealed a linear dependence of iPP domains size on EP content. The diameter of iPP domains, considered as spherical, is decreased from about 2.4 μm for LDPE/iPP through ~1.9 and ~1.6 μm for LDPE/iPP/EP-5 and LDPE/iPP/EP-10, respectively and to ~1.1 μm for LDPE/iPP/EP-15. Moreover the particle size distribution was narrower for high EP content. These SEM observations indicate clearly that ethylene-propylene copolymer acts as compatibilizer for LDPE/iPP blends and the observed morphological variations correlate well with the tendencies of the ternary blends mechanical properties. In the case of semi-crystalline polymer based blends, addition of ethylene/propylene rubber is expected to decrease the yield strength and modulus while increasing the elongation at break. [5] This assumption was confirmed to a large extend by our experimental result presented in Figure 5 where the changes in the young modulus and tensile and elongation at break are presented as functions of the EP content. For the EP composition range investigated in this study, tensile at break is a linear function while improvement in the elongation at break is observed from a EP content of 5 wt.-%. Above this threshold EP content, ε_b exhibits a nearly linear increase and elongation at break of about 850% are obtained. This can be interpreted in terms of enhanced compatibility between the amorphous phases of LDPE and iPP.

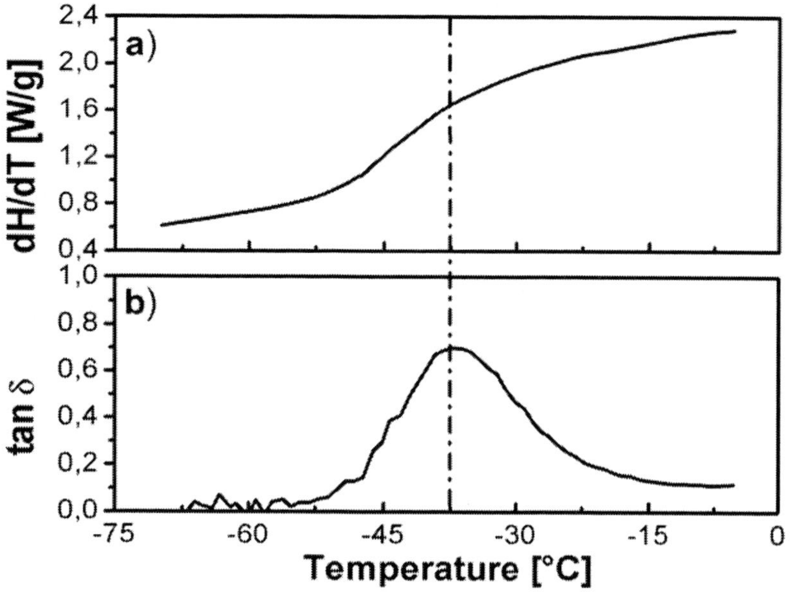

Figure 2. Temperature dependences of a) the heat flow recorded with the rate 10°C/min and b) the mechanical loss factor (tan *δ*) recorded during heating with the rate 2°C/min showing the rubbery character of the ethylene-propylene copolymer.

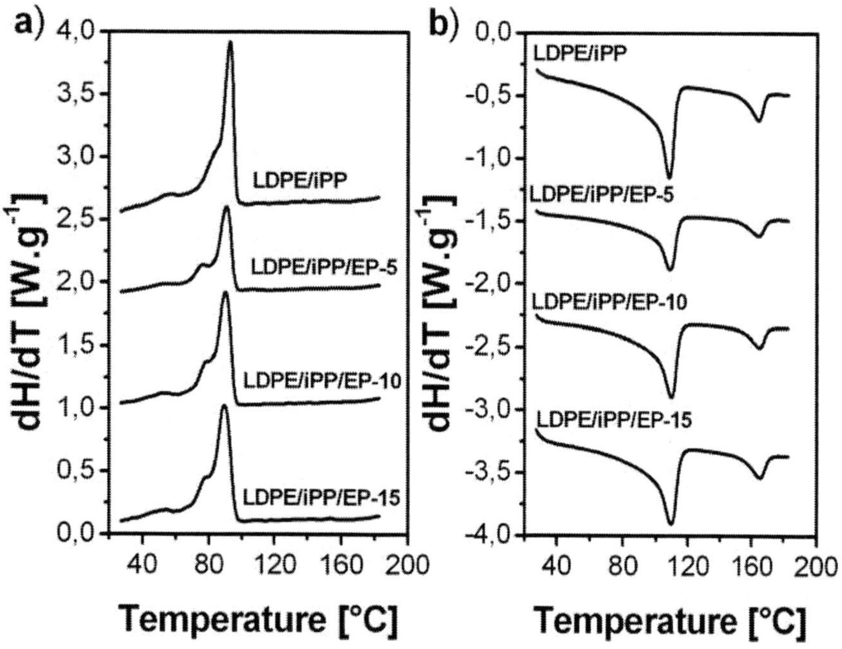

Figure 3. DSC thermograms of the binary LDPE/iPP blend and ternary LDPE/iPP/EP-*x* blends with three EP contents (5, 10 and 15 wt-%) recorded during a) the cooling from the melt and b) the subsequent heating.

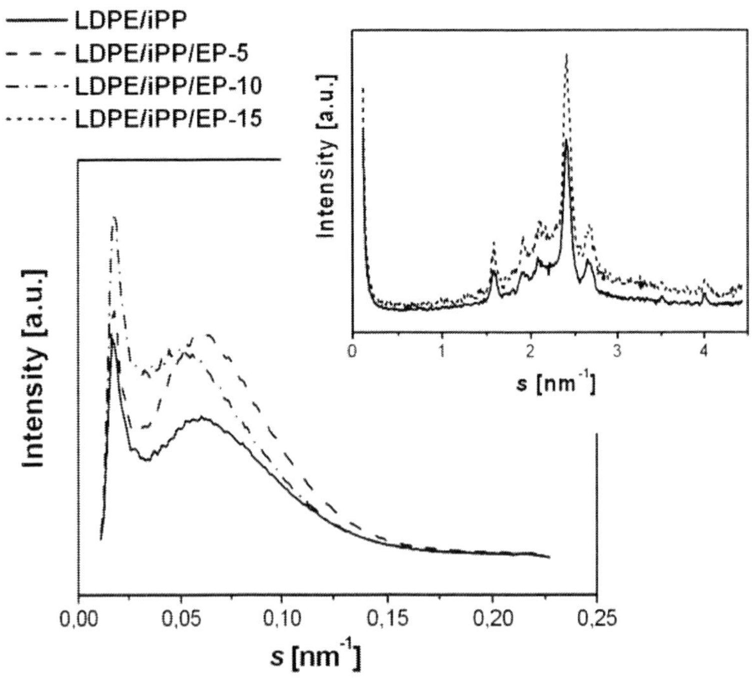

Figure 4. SAXS profiles recorded for LDPE/iPP, LDPE/iPP/EP-5 and LDPE/iPP/EP-10 blends. The inset presents the WAXS profiles recorded for LDPE/iPP and LDPE/iPP/EP-15 blends. The presented scattered intensity distributions were recorded at room temperature. They are presented as functions of the scattering vector ($s=2\sin\theta/\lambda$, where 2θ is the scattering angle).

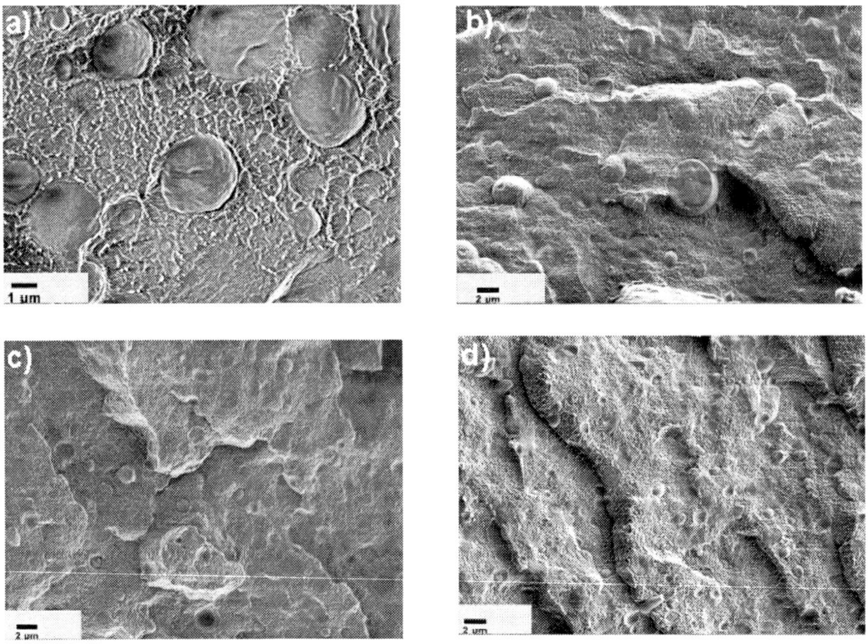

Figure 5. SEM pictures of fractured samples showing the phase dispersed morphology for a) LDPE/iPP, b) LDPE/iPP/EP-5, c) LDPE/iPP/EP-10 and b) LDPE/iPP/EP-15 blends.

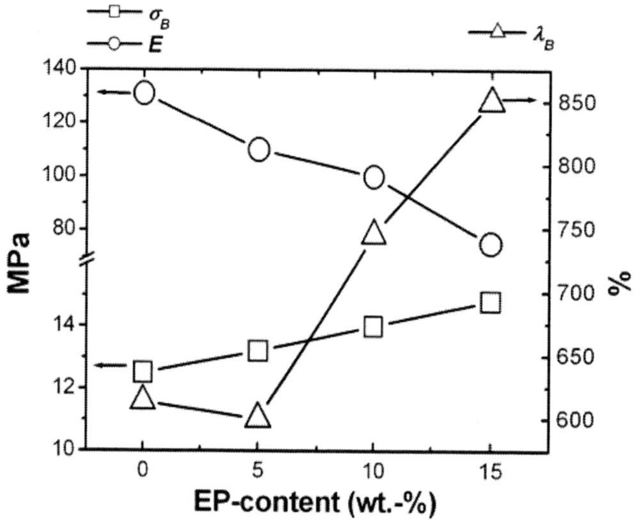

Figure 6. Dependences of tensile and elongation at break and young modulus for ternary LDPE/iPP/EP-*x* blends as functions of EP content. Meaning of the determined parameters is illustrated in Figure 1.

Conclusion

In this communication, we studied the effect of adding an amorphous ethylene-propylene copolymer to a blend of low-density polyethylene and isotactic polypropylene (80/20). In the EP composition range investigated, a significant compatibilization effect was observed as evidenced by morphological observation and mechanical tests.

References

[1] Utracki, L.A. *Polymer Alloys and Blends*; Hanser : Munich, Germany, 1989, p. 92. b) Paul, D.R. and Newman, S.M. *Polymer Blends*; Ed. New York: Academic, vol 2, 1978, p.319. c) Rajasekaran, J. J., Curro, J. G., Honeycutt, J. D. *Macromolecules*, 28, 6843, 1995. d) Lo, C-T., Seifert, S., Thiyagarajan, P., Narasimhan, B. *Polymer*, 45, 3671, 2004.

[2] Li, J., Shanks, R. A., Long, Y. J. *Appl. Polym. Sci.*, 82, 628, 2001. b) Li, J., Shanks, R. A., Long, Y. J. *Appl. Polym. Sci*, 87, 1179, 2003.

[3] Barentsen, W.M., Heikens, D. *Polymer,* 14, 579, 1974. b) Xu, G., Lin, S. Polymer, 37, 421, 1996. c) Caporaso, L., Iudici N, Oliva L. *Macromolecules*, 38, 4894, 2005.

[4] Manaure, A. C., Morales, R. A., Sanchez, J. J., Müller, A. J. J. *Appl. Polym. Sci.*, 66, 2481, 1997. b) Arnal, M. L., Matos, M. E., Morales, R. A., Santana, O. O., Müller, A. J. *Macromol. Chem. Phys.*, 199, 2275, 1998.

[5] Robertson, R. E., Paul, D. R. J. *Appl. Polym. Sci.*, 17, 2579, 1973.

In: New Trends in Biochemical Physics Research
Editors: S. D. Varfolomeev et al., pp. 9-21
ISBN 1-60021-463-0

Chapter 2

Analysis of the Historical Dynamics of Differences in Lifespan of Men and Women in Economically Developed Countries

V. B. Mamae and A. A. Tsarin
N. M. Emanuel Institute of Biochemical Physics, Russian Academy of Sciences
Moscow, Russia

Abstract

The historical dynamics of men's and women's lifespan differences is analyzed. Regular changes of men's overmortality during different historical periods are found. In a number of developed countries significant growth of men's overmortality with subsequent decrease during the period 1960 through 2000 is found. The start and the end of the men's overmortality peak differ in different countries. In some countries the decrease of the men's overmortality has not begun yet, but the preceding dynamics coincides with the dynamics in the other countries, which allows predicting the changes of the men's mortality in the near future.

Keywords: ageing, mortality, death rate, population, gerontology, lifespan, men, women, historical trends

Introduction

Analysis of the sexual difference in the lifespan and mortality distribution always attracted attention. Indeed, two almost identical organisms living in the same conditions and having small biological differences, often respond quite differently to the same influence. The fact that the fundamental biochemical differences are well studied suggests an idea about the possibility of revealing biological mechanisms and sense of the differences in the dynamics of the mortality.

In this article, the dynamics of sexual differences of the mortality is researched in detail for the populations of Finland and Sweden. It is also examined how universal the tendency of the growth of the men's overmortality in the latest half of the century is for the populations of the following countries: Denmark, England, France, Italy, Japan, Netherlands, Norway, Spain, Switzerland, USA.

Materials and Methods

The mortality data were obtained from The Human Mortality Database (http://www.mortality.org). The countries and historical periods that are researched are listed in the Table 1. The age dependence of the mortality is approximated in the age interval 10 through 84 years with the Gompertz-Makeham equation (1):

$$m_x = C+Re^{Ax} \quad (1)$$

where x – age, m_x – death rate in the age x, C – background component of the mortality, Re^{Ax} – age-related component of the mortality, R and A – parameters of the age-related mortality, as described before [1].

Table 1. Countries and historical periods.

Country	Years
England	1841-1998
Denmark	1921-2000
Spain	1908-2002
Italy	1872-2001
Netherlands	1850-2003
Norway	1846-2002
USA	1959-1999
Finland	1878-2002
France	1899-2002
Switzerland	1876-2002
Sweden	1751-2002
Japan	1950-1999

Results and Discussion

Sexual Dimorphism of the Age-related Mortality

Since Darwin, the male's overmortality is considered as a natural and constitutional property determined only by sex [2, 3, 4]. Nevertheless the analysis of the historical dynamics of the age-related mortality from 1751 through 2002 in Sweden and from 1878 through 2002 in Finland shows that the value of the men's overmortality undergoes regular changes (Figures 1, 2).

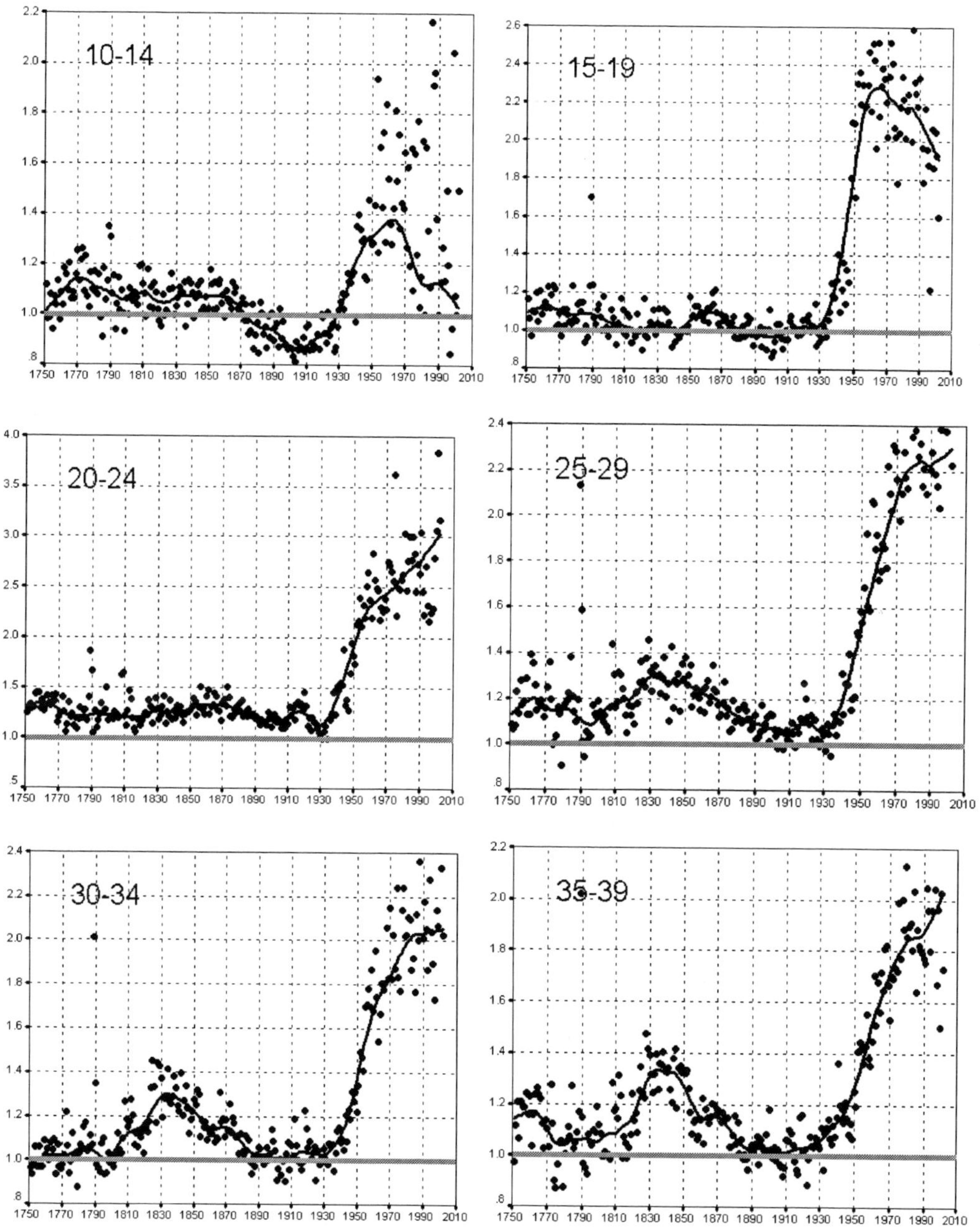

Figure 1. (Continued)

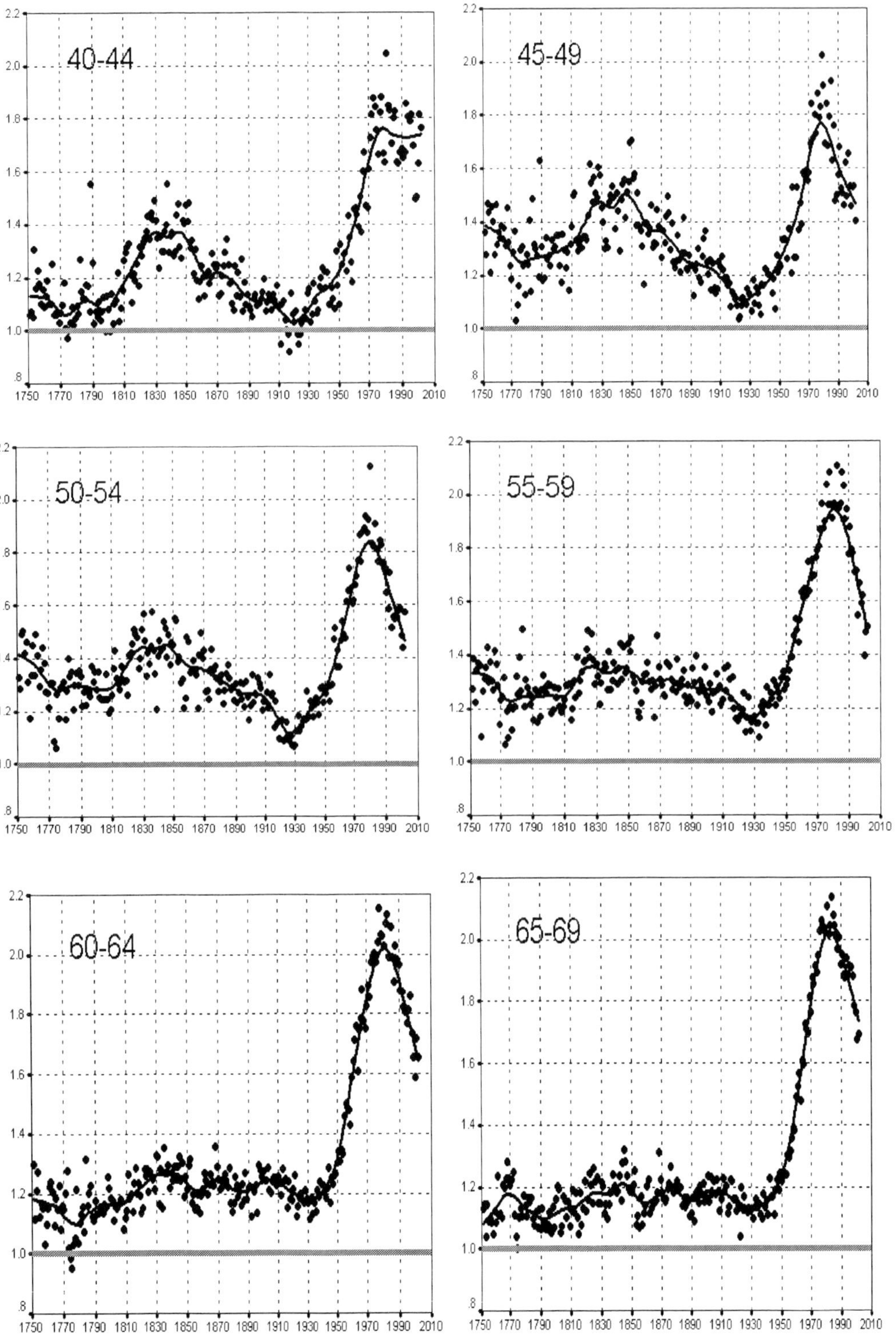

Figure 1. (Continued)

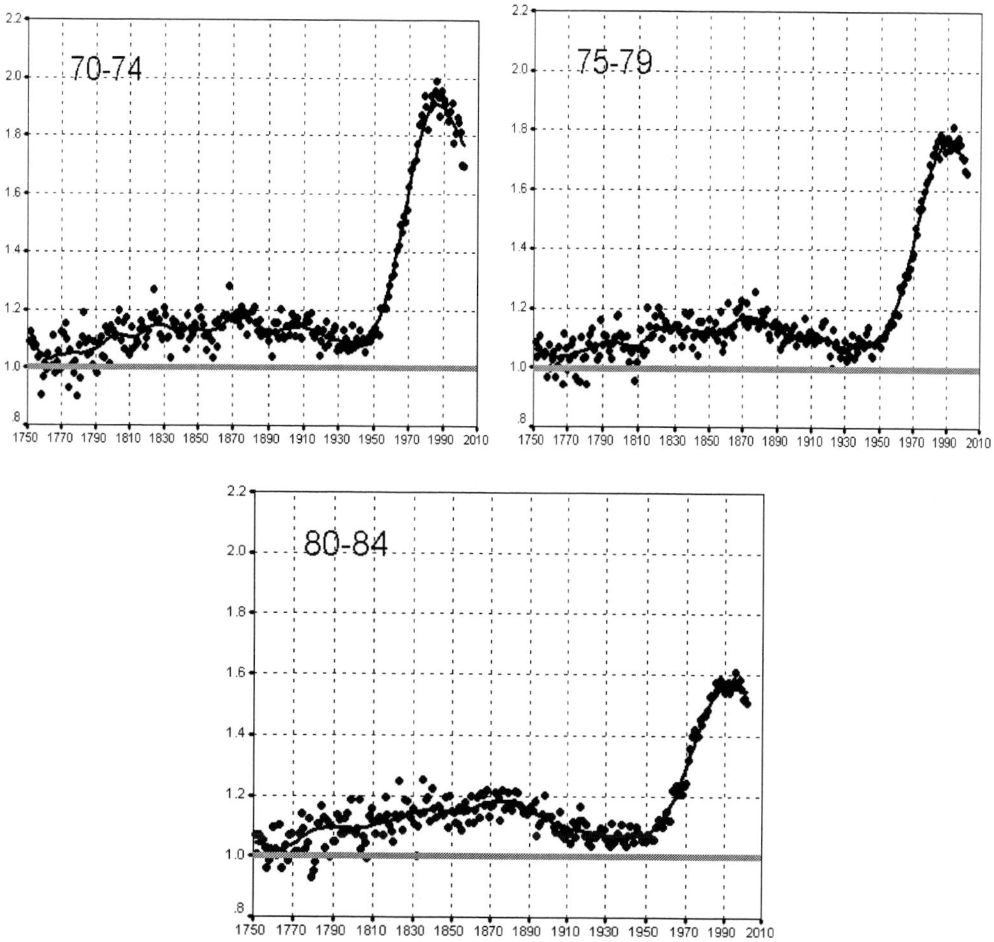

Figure 1. The historical dynamics of the ratio of the men's death rate to the women's death rate for the different age groups of the Swedish population from 1751 through 2002.

Before analyzing tendencies of overmortality changes, let us pay attention to the pronounced peaks. In Sweden in 1789 in the age groups 15 through 50 years high overmortality of men is observed. This peak is not present in younger and older age groups. Obviously, this peak is connected to a war that resulted in high mortality of men of average military age (Russian-Swedish War of 1788-1790).

In the historical dynamics of the mortality of the men and women (figures are not presented in this article) other mortality peaks are also visible, but since they are equal for men and women those peaks are absent on the Figure 1. In all the age groups in 1773 and in 1809 peaks are observed (Russian-Swedish War of 1809). Up to 50 years a peak is observed in 1918 ("Spanish disease", Finnish Civil War). Up to 20 years a peak is observed in 1857.

During further analysis of historical tendencies, we do not take into account the listed above peaks.

During the historical period concerned it's possible to separate 7 characteristic intervals. Their limits are slightly different for different age groups, sometimes up to 5-10 years. That's why the specified years are approximate.

1. 1751-1800. In all the age groups the ratio of the men's death rate to the women's death rate is nearly constant and has values near 1 (the mean values are presented in the Table 2).
2. 1800-1830. In the age groups from 30 up to 55 years increase of the men's overmortality is observed.
3. 1830-1850. In the age groups from 30 up to 55 years relatively high men's overmortality is observed.
4. 1850-1890. In the age groups from 30 up to 55 years the men's overmortality decreases to the former values.
5. 1890-1935. Low men's overmortality in all the age groups. Sometimes men's mortality is even lower than women's mortality (see Table 2).
6. 1935-1980. Increase of the men's overmortality in all the age groups is observed. The specified limits are rather conditional, because the older is the age group, the later the period starts and ends. In the age group 10-15 years the limits of this period are 1935-1950, whereas in the age group 80-85 years the limits are 1960-1990.
7. 1980-2002. In this period, starting from 45 years the men's overmortality significantly decreases. In the younger age groups the dynamics differs. It slightly decreases in some groups, whereas in other groups it remains at the reached value. In the age group 30-35 years the men's overmortality even increases. In the young groups the fluctuation of the values increases from year to year. This can be explained by the low level of the mortality, which results in high variations of the ratio of the men's death rate to the women's death rate.

Table 2. The value of the men's overmortality in Sweden during the stable periods.

Age	1751-1800	1830-1850	1890-1935
	Mean(m_M/m_W)	Mean(m_M/m_W)	Mean(m_M/m_W)
10-14	1.09±0.01	1.06±0.01	0.92±0.01
15-19	1.09±0.01	1.02±0.01	1.00±0.01
20-24	1.27±0.02	1.25±0.01	1.18±0.01
25-29	1.15±0.02	1.28±0.02	1.07±0.01
30-34	1.02±0.01	1.27±0.02	1.03±0.01
35-39	1.10±0.02	1.32±0.01	1.03±0.01
40-44	1.10±0.01	1.37±0.02	1.08±0.01
45-49	1.30±0.01	1.47±0.03	1.17±0.01
50-54	1.31±0.01	1.43±0.02	1.20±0.01
55-59	1.27±0.01	1.34±0.01	1.24±0.01
60-64	1.15±0.01	1.26±0.01	1.21±0.01
65-69	1.13±0.01	1.20±0.01	1.16±0.01
70-74	1.05±0.01	1.13±0.01	1.12±0.01
75-79	1.05±0.01	1.13±0.01	1.10±0.01
80-84	1.06±0.01	1.15±0.01	1.10±0.01

In Finland (Figure 2), as well as in Sweden, in some years high peaks of the men's overmortality are observed. In 1918 (Finnish Civil War) the men's overmortality is especially high in the age groups 15 through 60 years. In the other groups this peak is also observed, but it's not so high. In 1940-1941 the high overmortality is observed in the groups 15-45 years, and in 1944 the high overmortality is observed in the groups 15-40 years (1940-1944 – periodical wars against USSR).

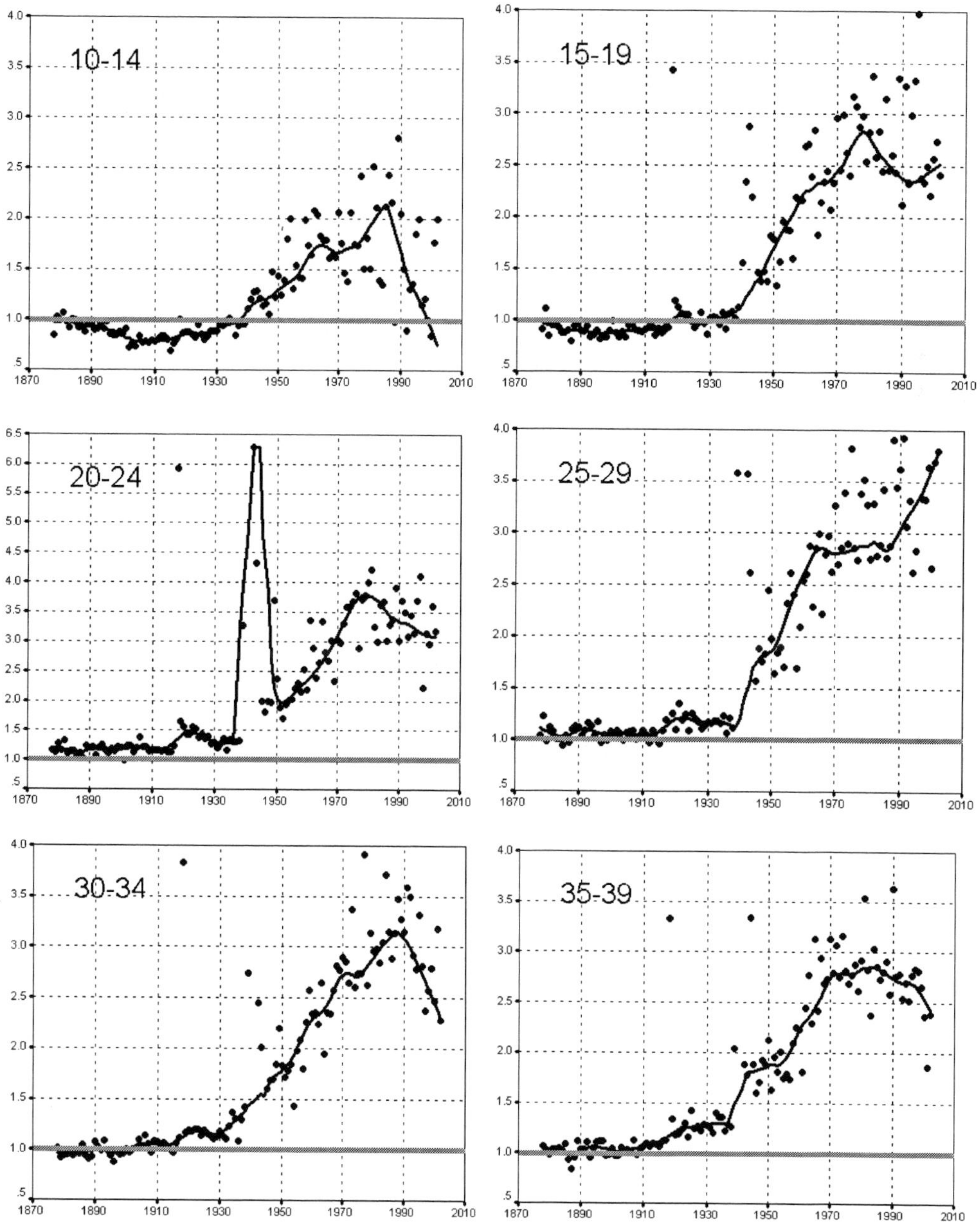

Figure 2. (Continued)

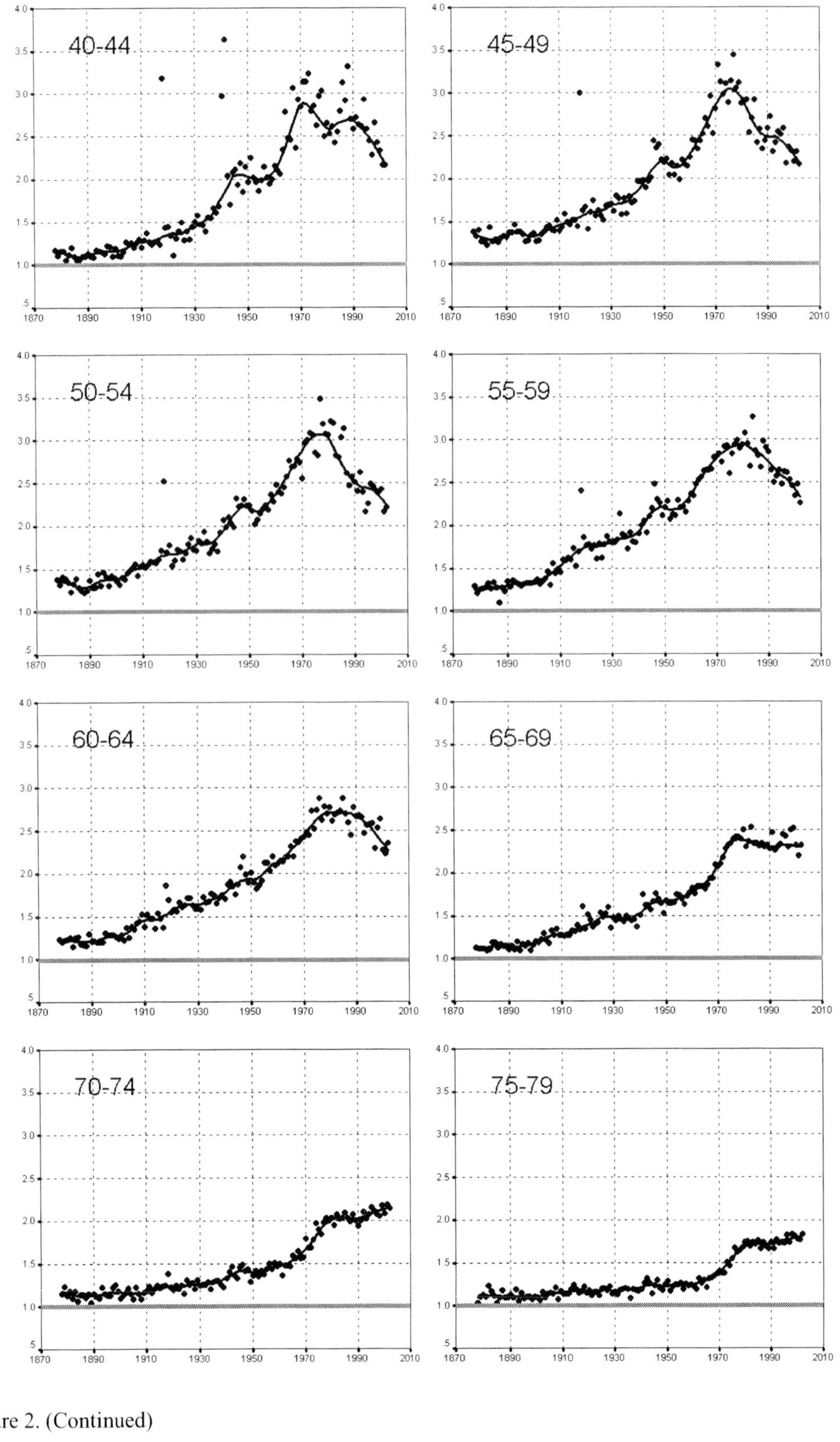

Figure 2. (Continued)

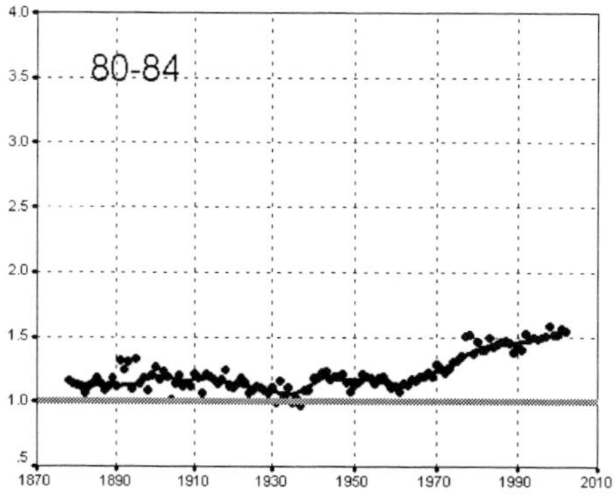

Figure 2. The historical dynamics of the ratio of the men's death rate to the women's death rate for the different age groups of the Finnish population from 1878 through 2002.

The listed peaks are not taken in the account further in the article.

During the overall interval being researched a single rising tendency of the men's overmortality up to 1970-1980 years is ovserved. The maximum is reached in different age groups in different years. Against the background of this global tendency more subtle effects are observed, which depend on age group.

The overmortality in the groups 10-40 years remains at the low level up to 1940 and sometimes even decreases below 1.

In the groups 10-20 years fluctuations of the men's overmortality significantly increases, which can be explained by the low level of the mortality in these groups.

In the groups 20-40 years after 1980 the overmortality remains at the reached level.

In the groups 40-65 years after 1980 noticeable decrease of the overmortality is observed.

In the groups 65-85 years after 1980 the overmortality either remains at the same level or continues to grows but with a lower rate then before 1980.

Table 3. The value of the men's overmortality in Finland during the stable periods.

Age	1878-1917	1919-1938	1970-1985
	Mean(m_M/m_W)	Mean(m_M/m_W)	Mean(m_M/m_W)
10-14	0.87±0.01	0.90±0.01	1.81±0.09
15-19	0.90±0.01	1.03±0.02	2.83±0.07
20-24	1.17±0.01	1.38±0.03	3.51±0.09
25-29	1.06±0.01	1.18±0.02	3.11±0.09
30-34	1.00±0.01	1.20±0.02	3.02±0.09
35-39	1.04±0.01	1.27±0.02	2.89±0.07
40-44	1.18±0.01	1.44±0.03	2.80±0.06
45-49	1.37±0.01	1.65±0.02	2.99±0.06
50-54	1.42±0.02	1.74±0.02	3.02±0.05
55-59	1.38±0.02	1.81±0.02	2.89±0.04
60-64	1.30±0.02	1.65±0.01	2.67±0.04
65-69	1.20±0.01	1.46±0.01	2.33±0.03
70-74	1.16±0.01	1.25±0.01	1.93±0.03
75-79	1.13±0.01	1.17±0.01	1.60±0.03
80-84	1.16±0.01	1.09±0.01	1.38±0.02

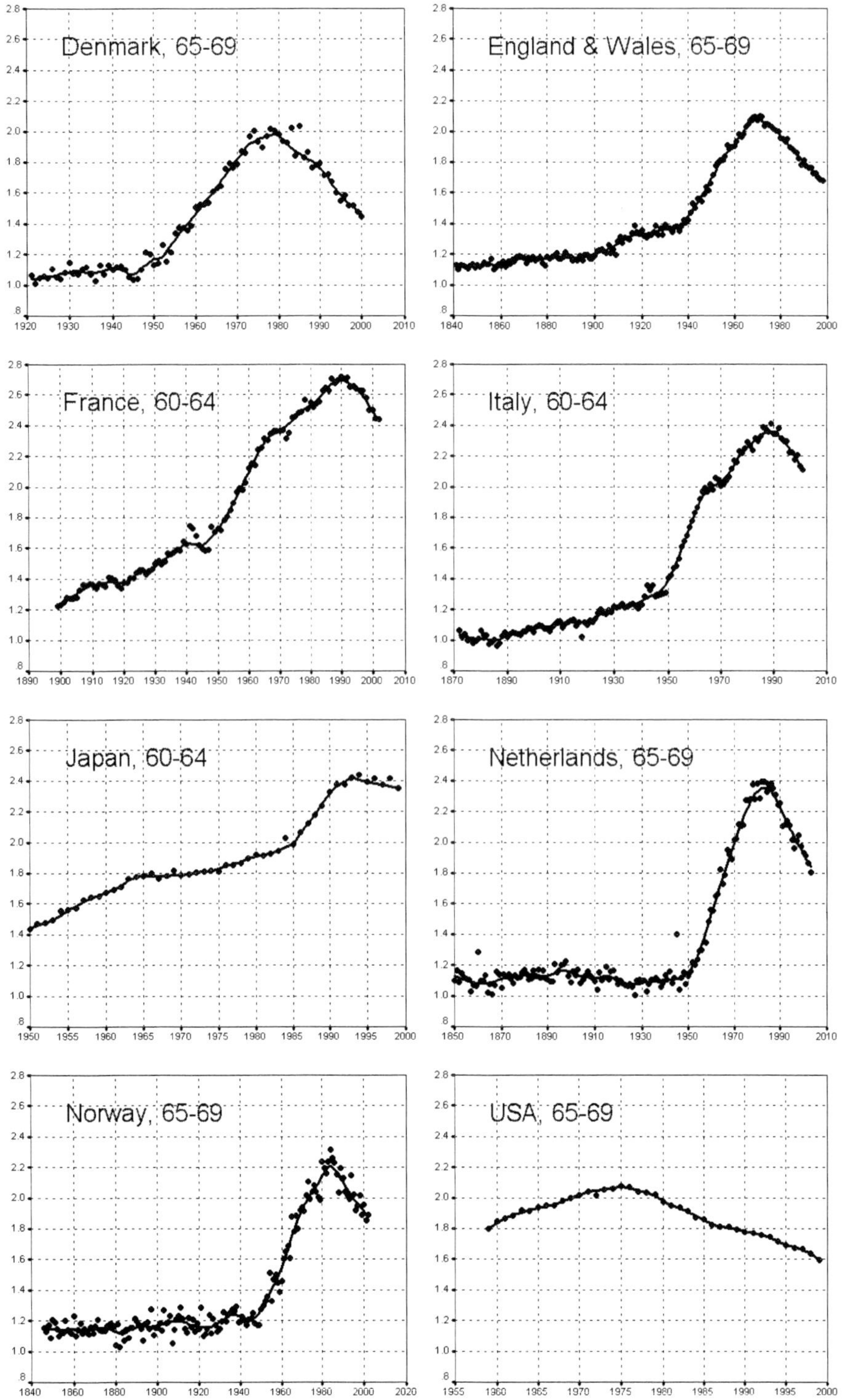

Figure 3. The historical dynamics of the ratio of the men's death rate to the women's death rate for some age groups of populations of the developed countries.

In Finland and in Sweden the common behavior is evident, which consist in the growth of the men's overmortality to 1980 and the subsequent decrease. In order to find out how global this tendency is, the statistical data of the following countries were additionally analyzed: Denmark, England, France, Italy, Japan, Netherlands, Norway, Spain, Switzerland, USA (Figure 3).

It turned out that the growth of the men's overmortality with subsequent decrease in the range 1960 through 2000 years is observed in all the analyzed countries. The limits and the position of the maximum differ in different countries. The value of the ratio of the men's death rate to the women's death rate is also different in different countries. The delay of the effect in the old age groups is observed in all the countries.

In some of the countries the growth of the ratio of the men's death rate to the women's death rate takes place simultaneously with decrease of both the men's and women's mortality. The ratio grows due to the delay of the changes of the men's mortality. Nevertheless, in the most countries in the specified period just the growth of the men's mortality takes place.

SEXUAL DIMORPHISM IN THE DYNAMICS OF THE PARAMETERS LNR AND A

The well-known fact of the high mortality of men appears in the coordinates lnR-A in the location of the curve (рис. 4). Women have higher values of the parameter A and lower values of the parameter lnR. Respectively, men's values have the reverse behaviour. This corresponds to the conception that at the high starting mortality (connected with lnR) the rate of its increase (i.e. the parameter A) must be lower.

What is not so evident is the sequence of changes of the curve. It's well visible at the Figure 4 that in the case of women the curve goes to a new stage from the right bottom part, whereas in the case of men the new stage starts from the left top part. An analogous situation is observed in Finland as well. It seems to us that the most logical explanation of this effect is influence of cohorts of high mortality. Passage of a generation of high mortality through the top limit of the interval of approximation can result in «regress» of the curve.

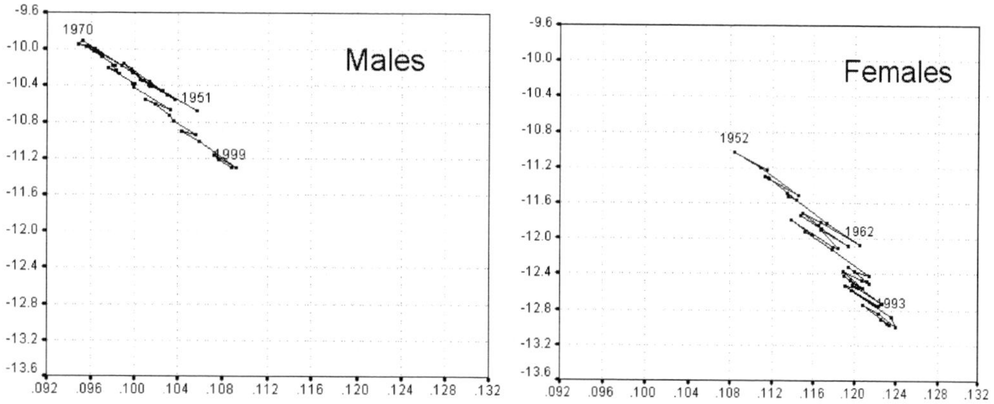

Figure 4. Dependence of lnR from A for the Swedish population.

Another feature of the dynamics of the parameters lnR and A consists in delay of changes in the case of men compared to the case of women (Figure 5). The value lnR starts decreasing for women and only after 30-40 years it starts decreasing for men.

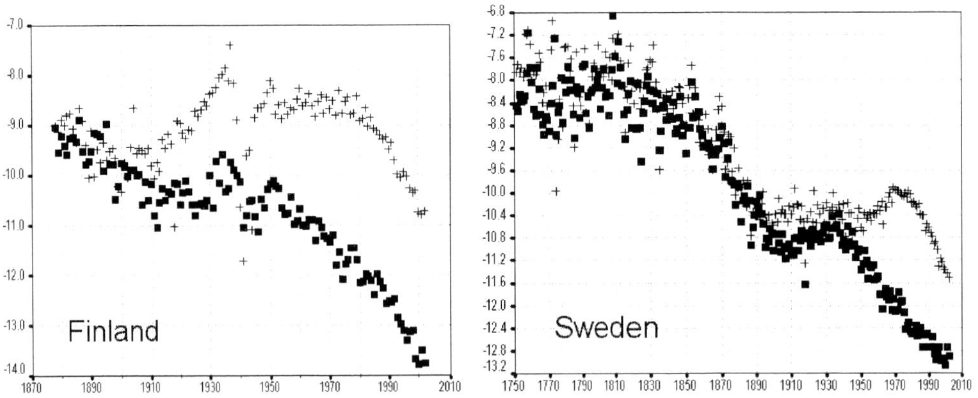

Figure 5. The historical dynamics of lnR. Sweden 1751-2002. Finland 1878-2002. Men (+) and women (■).

Conclusion

1. The ratio of men's and women's mortality undergoes regular changes in the historical dynamics, which appears as follows:
   a. It's possible to separate intervals, inside of which the men's overmortality fluctuates near a value with transitional periods between those intervals.
   b. The value of the men's overmortality in different age groups obeys to a single regularity. Nevertheless it has age-related features.
2. A single tendency is observed in many developed countries:
   a. There is growth of the men's overmortality in the period 1960-2000 with subsequent decrese. The limits of the period and the location of the maximum differ in different countries. Nevertheless, the decrease of the men's overmortality does not start yet in the old age groups in some countries.
   b. The maximum is observed later in older age groups.
3. The decrease of mortality of men in the latest half of the century is delayed in comparison with the decrease of the mortality of women.

References

[1] V.B. Mamaev, A.A. Tsarin, E.A. Minenkova. The historical dynamics of the age-related mortality. 1. The evolution of the specific lifespan of the Finnish population. *Adv. Gerontology,* Vol.15, pp.23-29, 2004.

[2] Ch. Darwin. The human origin and the sexual selection. *Collected works.* Vol.5. Moscow, USSR AS, 1953.

[3] *Comfort. The biology of ageing.* Moscow, Mir, 1967.
[4] L.A. Gavrilov, N.S. Gavrilova. *The biology of the lifespan.* Moscow, Nauka, 1991.

In: New Trends in Biochemical Physics Research
Editors: S. D. Varfolomeev et al., pp. 23-30
ISBN 1-60021-463-0

Chapter 3

SURFACE MODIFICATION OF POLYPROPYLEN FIBRES BY OZONE WITH THE PURPOSE OF GIVING THEM THE NEW CONSUMER PROPERTIES

O. V. Alekseeva, M .L. Konstantinova and S. D. Razumovskij
Institute of Biochemical Physics, N. M. Emanuel
The Russian Academy of Sciences, Moscow, Russia

At contact of ozone with polypropylene fiber (PP) ozone quickly enough reacts with it. Reaction proceeds in surface nanolayer (10^{-9}м), thus many properties of a fiber essentially change. The complex of experimental works in which changes of physical - chemical properties of PP fiber and nonwoven term bound polypropylene material received during its reaction with ozone-oxygen mix are investigated is executed. It is established, that reaction proceeds in surface layer of a fiber and is accompanied by substantial growth of the PP fibers module of elasticity at break (23 %), and also significant, increase in a moisture capacity.

Keywords: nanolayers, polypropylene, fibers, nonwoven material, ozone

INTRODUCTION

The general release of synthetic fibers and fabrics on their basis today exceeds 20 megaton/year and continues to increase [1]. Practically, all known types of chemical fibers and yarns, except for special kinds, can be essentially used as for technical, and house textile [2,3]. Significant share fibers make of let out assortment and fabrics on the PP base, and this share continues to increase annually. In particular, from PP fibers make floor carpets (carpet (BCF) strings) and nonwoven materials (spanbond/meltblaun) for air filters and other products (staple fiber). Properties of fibers diversify, mixing them with a clap, viscose, wool and other components [4,5,6]. Many ways of giving polypropylene materials different in comparison with initial PP of properties are known [7]. Fibers impregnate with antistatic, paint in different colors, extend with the purpose of giving of anisotropic properties. Among methods of chemical modification it is necessary to mention ozonization as very effective way of surface modification of products from PP.

The listed methods essentially expand as a scope of polymer, or create new. In turn expansion of sphere of application of synthetic materials creates preconditions for development of new technologies and, in particular, does our work actual.

The purpose of the present work was in:

- Research of process of polypropylene fibers surface modification.
- Measurement of physical - chemical characteristics of polypropylene fibers and nonwoven materials on the basis of polypropylene fibers under influence on their surface ozone

Below results of experimental researches are shown.

EXPERIMENTAL PART

As objects of our research the polypropylene fiber (PP) consisting of 30 fibers by thickness of 20 microns, with length 15 have been chosen. The linear density of a fiber made 0,91gr/sm^3. And nonwoven term bound material with superficial density 20gr/м^2 and thickness 8мм.

PP fiber or a nonwoven material place in a reactor and treat to action of the ozone-oxygen mix. Ozone receive by transmission of oxygen jets through a zone of the electric category (4-6 kV) with a speed of 100 ml/min. Concentration of ozone register on change of optical density of a gas stream at l=254 nanometer on an input and an output from a reactor. For processing fibers and a nonwoven material use of 2 types of reactors is necessary: one static - for studying bracket and are treated by ozone-oxygen mix at constant rotation which provides uniform influence of ozone on a surface of a polymeric fiber. On a difference in concentration of ozone on an input and an output from a reactor it is possible to predetermine depth and speed of reaction. Time of ozonization steals up empirically depending on substances size. In particular, for samples in weight 0,1-0,3 time of ozonization will make: ~3 - 9 minutes

Thickness of a bunch and separate fibers during processing by ozone is convenient for supervising by means of a microscope " POLAM C-111 ". For this purpose in the beginning spend an initial estimation of thickness of the initial raw samples on a scale to a field of a microscope. Then fibers place in a reactor, process the ozone, demanded time, take from a reactor and again measure thickness of a fiber. It is in passing possible to observe of changes in a microstructure of a fiber and its transparency under action of ozone. The important characteristic of the majority of fibers is them hydrophily which in our case is consequence of radical changes in a microstructure of fibers under influence of ozone. A measure fibers hydrophily can be sorption of water. Using this property it is possible to define a modification degree of the sample (the express train an estimation) under influence of ozone. For measurement of material sorption fibers or a nonwoven material place in a vessel with water at a room temperature, take out, delete a superfluous moisture and again weigh. The attitude between indications of weights for moistened in water and the raw PP samples is a measure sorption abilities of a fiber. Weighing is spent at a room temperature.

The structure of functional groups and degree of PP material modification estimate according to infra-red spectroscopy IR spectra of fibers and strings of a nonwoven material from PP register on infra-red spectrophotometer, Specord-71 JR, by means of prefixes MNPVO. Depth of penetration of radiation in the sample makes ~0,5 and 0,9 microns for prisms from Germany (a corner of falling Θ =55, N=20) and КЗС-5 (Θ =55, N = 12) accordingly. Samples before registration of spectra compact between the polished steel plates that increased the area of their contact to reflecting sides of prisms MNPVO.

Measurement of a fiber durability, before and after ozone influence on its surface make on a stretching of PP fibers. Measurements are spent at a room temperature by test machine INSTRON with digital record of the loading diagram. Speeds of loading at any kind of tests remain constants and make $5*10^{-4}$ km/s. The processed and raw fibers are cut on samples in length of 0,17 m. Length of a working part made 0,15 m. Durability of investigated fibers at destruction δ_p, their initial modules of elasticity E and limiting deformations $\varepsilon_к$, count from loading diagrams. Durability at break δ count under the formula:

$$\delta = P:S$$

Where P - loading, S - the area of cross-section section of a fibers bunch; the module of elasticity E define by a linear part of the stretching diagram of a fibers bunch :

$$\delta = f(\varepsilon).$$

Results and Discussion

On Figure 1 dependences of ozone concentration change on an input and an output for an empty reactor (1) and the sample of PP fiber containing (2) from time are presented. The areas limited by axes of ordinates, absciss and experimental curves are equivalent to the charge of ozone on filling of a reactor and on reaction with the sample.

The area limited by curves 1 and 2 (it is shaded in figure) is equivalent to the ozone charge on the reacted substance. From figure it is possible to see, that the material absorbs appreciable quantity of ozone (in our case is $2{,}14\cdot10^{-5}$ mol O_3 per 1gr. of substance). Absorption proceeds quickly, that, apparently, it speaks high reactionary ability of the active centers on the surface in relation to ozone [9]. After ~3 minutes absorption of ozone stopped. The difference between the areas above curves 1 and 2 is equivalent to quantity of the absorbed ozone. The estimation of a parity of the absorbed ozone to weight of the sample has shown, that 3-5 monolayers from surface layer participated in reaction ($2{,}4\sim2{,}6\cdot10^{-9}$м). The stop of reaction is caused by an expenditure of is reactionary-capable groups in this layer, apparently, as a result of diffusion braking.

From data on ozone absorption by PP the quantities of ozone spent for reaction with weight shares of polypropylene at different depths of transformation surface layers of PP have been calculated. It is visible, that in a wide interval of transformations good enough linear correlation is kept: for 1 monomer part of PP circuit it is spent about 1 mol. ozone.

On the basis of the aforesaid it is possible to draw a conclusion that reaction goes basically on a surface of the sample.

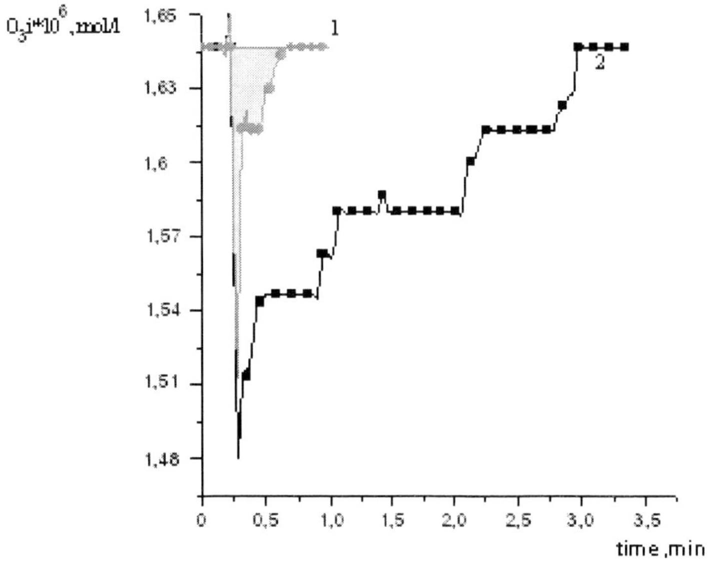

Figure 1. Dependence of concentration change of ozone $[O_3]_i$ on an input and an output from a reactor for an empty reactor (1) and containing the sample of PP fibre (2) from time

As acknowledgement it is possible to cite data IR -spectra on structure of functional groups and research of labile intermediate products by method EPR. Method IR-spectroscopy preliminary investigates structure of products of reaction polypropylene fibers with ozone (received on a difference of IR-spectra of initial polymer and the product obtained after ozonization) (Figure 2).

Figure 2. IR - a spectrum of products structure of polypropylene fibres reaction with ozone

The maximum of absorption at 1740-1760см-1 in IR - a spectrum is caused by fluctuations C=O-group that is characteristic for aldehydes and complex aliphatic ethers. Reduction of number of the resolved strips and the general increase in intensity of absorption speak about downturn of symmetry of molecules of ozonized PP.

During ozone influence fiber or a nonwoven material undergo essential changes. These changes are convenient for estimating as visually, and instrument methods.

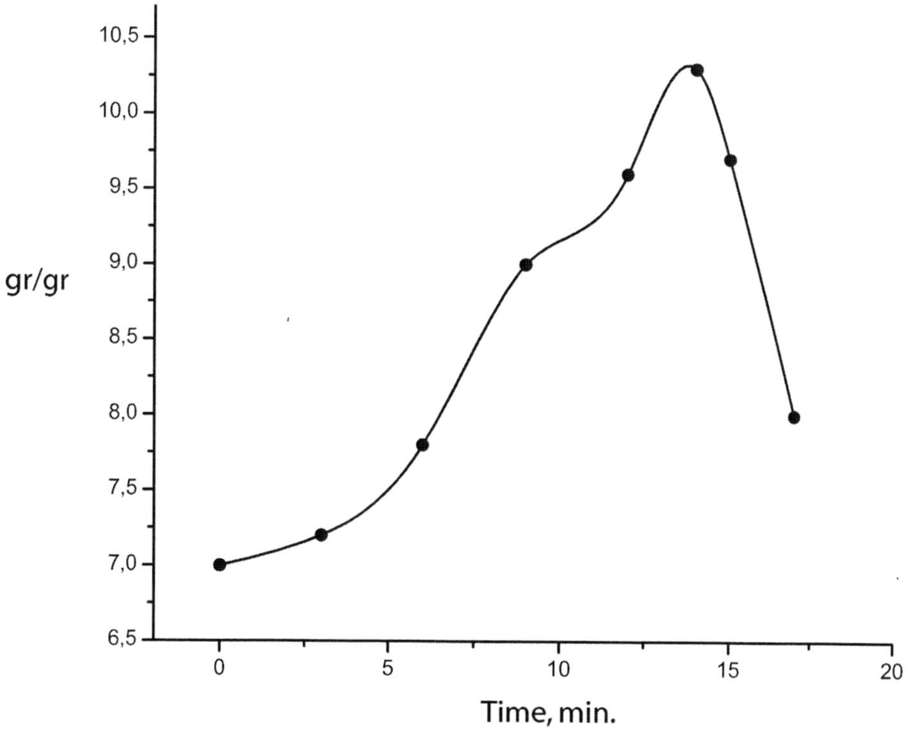

Figure 3. Change of water sorption of PP fibre as function of ozone action

Action of ozone on a surface of polymeric fibers is usual accompanied by their sorption characteristics change [8]. Polypropylene is not exception. On Figure 3 shown the waters sorption change data by a PP fiber, as functions from action of ozone. It is possible to see, that in the beginning growth of sorption, and then its sharp recession is observed during ozonization. Apparently, in the beginning, at small quantity of the absorbed ozone, formed in surface layer functional groups which has lyophilic nature, at the further action of ozone they turn to the others, more waterproof. Similar changes in sorption characteristics are also observed at ozone processing to nonwoven material on the basis of PP fiber (Figures 4,5). Concentration of ozone for modification a nonwoven material was $1{,}93 \cdot 10^{-4}$ mol O_3. Using this property it is possible to define a degree of updating of the sample under influence of ozone. The termination of influence at this or that stage of processing will allow to receive a material with the set and necessary properties, i.e. to pick up a mode of processing.

Essential changes of a material surface (thickness of a bunch and separate fibers during processing by ozone supervised by means of a microscope to "POLAM C-111") are appreciable. The surface becomes less defective and more smooth during processing by ozone. Also the transparency on 30 % (optical transmission) changes. Bleaching an initial PP material during processing by ozone is observed. The fibers representing extended crystallity, are in addition split on ozone action on even more thin. Action of ozone on a nonwoven material leads to that the material fluff up and increases in thickness (with 8мм up to 12мм) approximately after 6 minutes from the beginning of processing by its ozone. Ozone action is not accompanied appreciable material destruction.

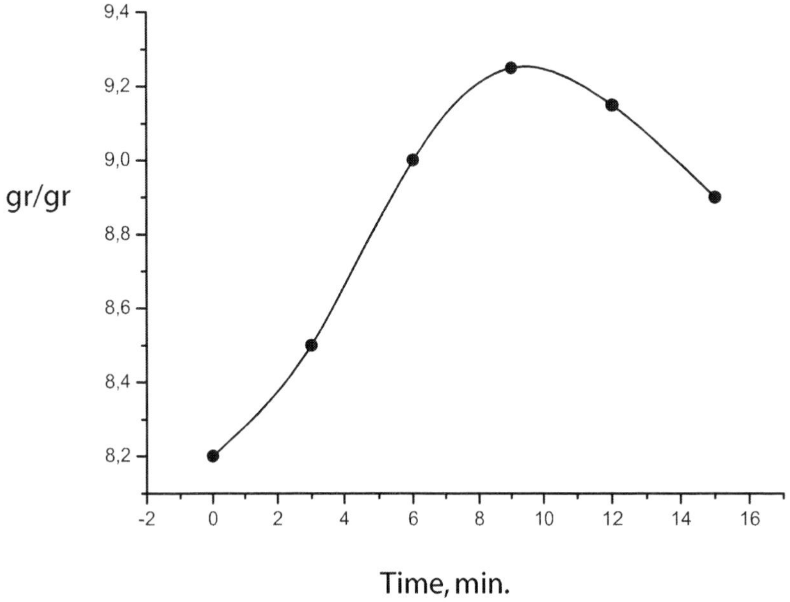

Figure 4. Change of water sorption of PP white nonwoven material as function of action of ozone

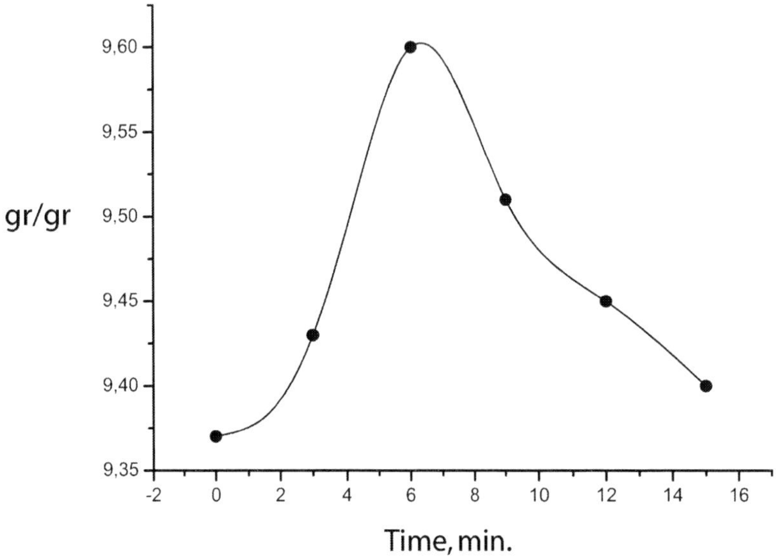

Figure 5. Change of water sorbtion of PP grey nonwoven material (a mix of software and a polyether 1:1) as function of ozone action

This changes in properties of a material are possibly caused by modification of polymer surface and formation of functional groups.

Have been investigated mechanical properties of PP fibers (tables №1, 2 and 3). At depth of penetration, a component a little nanosize layers (10^{-9}) [9], and small expenses of ozone appreciable effects in change of physical-mechanical properties of samples are received. From the given tables follows (Graf.6), that the module of elasticity varies not monotonously. The module of elasticity calculated as the attitude of a mechanical pressure to lengthening changes over a wide range. Thus dynamics of change has distinctly expressed extreme character. The maximal value of the module of elasticity (extremum) is reached at 260c. influences (~23 %).

Table 1. Dynamic of hardness change.

Strength (GPa)	0,58	0,8	1,38	0,98	3,4	1,5	0,9	0,5
Processing Time (s)	0	45	80	150	270	367	432	496

Table 2. Change of elongation of a sample in a dependence from time of processing.

Elongation(%)	20,4	21,6	23,2	26,7	29,7	29,8	26,2	24,4
Processing Time (s)	0	45	80	150	270	367	432	496

Table 3. Change of a modulus of elasticity of a filament from time of processing.

Modulus of elasticity (GPa)	7,6	10,4	18	12,8	32,8	10,3	8,9	8,1
Processing Time (s)	0	45	80	150	270	367	432	496

Ability of a fiber to be stretched (table.2), long enough remains without changes and only by the end of process increases for ~9 %. On a background of changes of other properties of a fiber this quality seems rather stable.

CONCLUSION

As follows from the given experiences ozone has appeared the effective modifying agent for products from polypropylene. Unlike the majority of other polymeric materials [10,11] complex of polypropylene properties under its influence improves. Its action is not accompanied appreciable material destruction. At insignificant depth of penetration (nanolayer) and small expenses of ozone greater effects in change of physical-mechanical properties of samples have been received. Reaction proceeds in surface layer of a fiber and during processing accompanied by change of waters sorption by a PP material. Using this property it is possible to define a degree of sample modification under ozone influence. The termination of influence at this or that stage of processing will allow to receive a material with the set and necessary consumer properties.

Also reaction is accompanied by substantial growth of initial value with the further recession of fibers module of elasticity for 23 %; elongation - on 9%. At the picked up modes of processing material destruction it is not observed.

Results of the present research allow to receive essential increase of qualitative characteristics of one of the most mass and industrially-important synthetic fibers and nonwoven materials on its basis, to expand spheres of its application and can be of interest for industrial realization.

ACKNOWLEDGEMENTS

In the conclusion authors express gratitude prof. Kuperman A. M. and dr. Solodilovu V. A. for the help in work on studying mechanical characteristics of software.

REFERENCES

[1] Aizenshtein E.M. / *Chemical fibers*, 2004, №4, c.60.

[2] Aizenshtein E.M./ *Chemistry and business*, 2004, № 5-6, c.50.

[3] Piller, Travnichek Z.*Sinteticheskie of a fibre and features of their processing in the textile industry.* Moscow, Izd. The foreign literature. 1960.

[4] Arbelaiz, G. Cantero, B. Fernández, I. Mondragon, P. Gañán, J.M. Kenny// Flax fiber surface modifications: Effects on fiber physico-mechanical and flax/polypropylene interface properties/ *Polymer Composites*, Volume 26, Issue 3, Date: June 2005, Pages: 324-332.

[5] Gonzalo Martínez-Barrera, Enrique Vigueras-Santiago, Susana Hernández-López, Gonzalo Martínez-Barrera, Witold Brostow, Carmina Menchaca-Campos// Mechanical improvement of concrete by irradiated polypropylene fibers/ *Polymer Engineering and Science*, Volume 45, Issue 10, Date: October 2005, Pages: 1426-1431.

[6] V. N. Hristov, ST. Vasileva, M. Krumova, R. Lach, G. H. Michler// Deformation mechanisms and mechanical properties of modified polypropylene/wood fiber composites/ *Polymer Composites*, Volume 25, Issue 5, Date: October 2004, Pages: 521-526.

[7] X. P. Zhou, R. K. Y. Li, X. L. Xie, S. C. Tjong //Reinforcement of polypropylene using sisal fibers grafted with poly(methyl methacrylate)/ *Journal of Applied Polymer Science.* Volume 88, Issue 4, Date: 25 April 2003, Pages: 1055-1064.

[8] Alekseeva O.V., Konstantinova M.L., Razumovskij S.D.//Dynamics of absorption of ozone polyamide films and fibres/*Polymer sciens,* A, 2004, №10, with. 1-4.

[9] Alexeeva, M.L.Konstantinova, S.D. Razumovskii, G.E. Zaikov. Physical-Mechanical Properties Change in polyamide Filament Under Ozone Influence. Chemical Physics Pyrolysis in Book "*Combustion and Oxidation*", Chapter 13, A.A. Berlin, I.A. Novakov, N.A. Khalturinskiy, G.E. Zaikov.

[10] D. J. Priest //Fold surface of polyethylene single crystals as assessed by selective degradation. I. Ozone degradation method/ *Journal of Polymer Science Part A-2: Polymer Physics*, Volume 9, Issue 10 , Pages 1777 - 1791, 2003.

[11] Sunil Dasgupta //Surface modification of polyolefins for hydrophilicity and bondability: Ozonization and grafting hydrophilic monomers on ozonized polyolefins/ *Journal of Applied Polymer Science/* Volume 41, Issue 1-2 , Pages 233 – 248.

In: New Trends in Biochemical Physics Research
Editors: S. D. Varfolomeev et al., pp. 31-35
ISBN 1-60021-463-0

Chapter 4

Thermodynamic Incompatibility of Locust Bean Gum and Methylcellulose in Water

Eva Richter and Susanne Gűnnel
Faculty of Food Chemistry, University of Jena, Jena, Germany
Yurij Antonov and Irina Plashchina
Institute of Biochemical Physics, Russian Academy of Sciences, Moscow, Russia

Abstract

Thermodynamic incompatibility is a fundamental physico-chemical phenomenon which determines the structure, rheology and functional properties of (bio)polymer mixtures [1]. Biopolymer mixtures are of importance to control the composition – structure-property relationship in natural biological systems and in formulated food. Therefore, to predict the properties of mixed biopolymer systems, their phase state and phase diagram should be known over a wide range of physico-chemical parameters [2].

The aim of the work is an investigation of the phase state and the phase behavior of the ternary water-methylcellulose (MC)-locust bean gum (LBG) system, determination and description of the phase diagram using the main thermodynamic parameters obtained.

Introduction

Thermodynamic incompatibility is a fundamental physico-chemical phenomenon which determines the structure, rheology and functional properties of (bio)polymer mixtures [1]. Biopolymer mixtures are of importance to control the composition –structure-property relationship in natural biological systems and in formulated food. Therefore, to predict the properties of mixed biopolymer systems, their phase state and phase diagram should be known over a wide range of physico-chemical parameters [2]. Most studies in this field relevant to the aqueous protein-polysaccharide systems [1]. The phase behavior of the ternary systems containing two linear polysaccharides which are important in industry is not well studied due to great difficulties in determination of the phase composition in such systems. This difficulty can be eliminated by using a simple and accurate the phase-volume method adapted from Koningsveld and Staverman [3], and Polyakov et al [4] allowing to determine the phase diagrams of the complex multicomponent systems without direct phase analysis. This method was developed at the Institute of Biochemical Physics, RAS.

The aim of the work is an investigation of the phase state and the phase behavior of the ternary water-methylcellulose (MC)-locust bean gum (LBG) system, determination and description of the phase diagram using the main thermodynamic parameters obtained.

Materials and Methods

Locust bean gum (LBG) is a linear consisting of β-(1,4)-D-mannose units. Approximately every fourth mannose units is substituted with a small side chain consisting of a 1,6 linked α-galactose sugar[5].

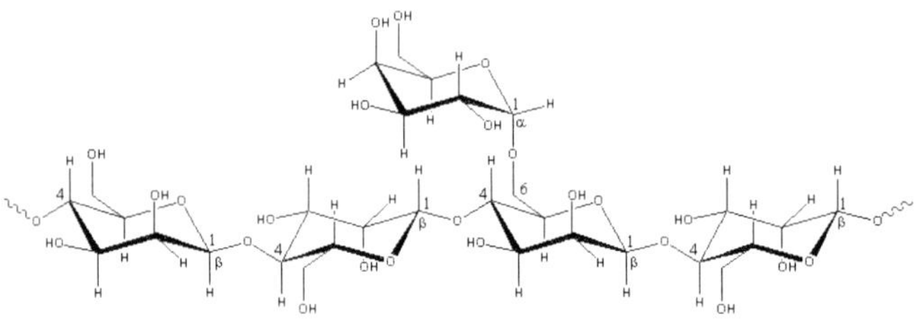

Figure 1. Molecular structure of LBG

The level of substitution has a major impact on the properties of the gums as polymannan can associate and self cross link. Hence ivory nut mannan can be completely insoluble whereas fenugreek shows the properties of a simple random coil. Locust bean gum shows intermediate properties in that it is soluble but shows substantial aggregation in solution.

LBG was purchased from Sigma Chemical Co. Its molecular weight (M_w) was approximately 1.1 10^6 Da.

Sample of methylcellulose (MC) was kindly obtained from System Bio-Industries. Methylcellulose is simple ether of cellulose. Structural feature of MC is presented in Figure 2. Molecular weight of MC used in present study was approximately 200,000 Da.

Both of these polysaccharides are widely used in industry.

The phase diagram of the ternary systems was determined at pH 5.0 and 298 K as illustrated schematically in Figure 3 a, b.

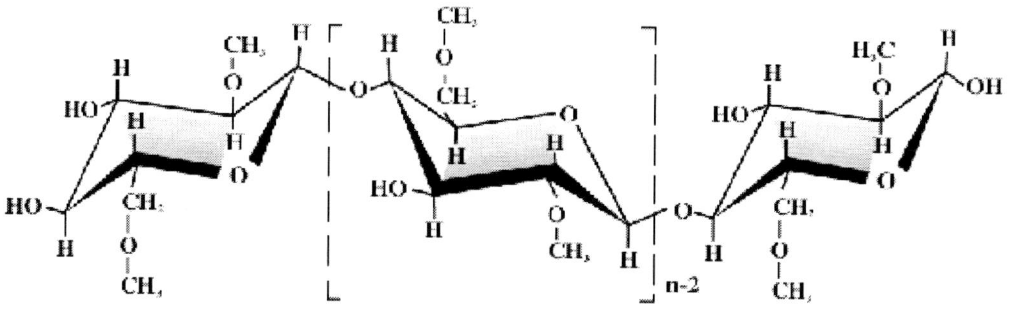

Figure 2. Structure of MC [5]

In these Cartesian co-ordinate diagrams, the horizontal axis corresponds to LBG concentrations, and the vertical one corresponds to MC concentrations. The procedure is adapted from Koningsveld and Staverman and Polyakov and colleges. The separately prepared LBG and MC solutions (concentrations C^o_2 and C^o_3, respectively) were mixed at 298 K in various weight ratios, yielding mixtures with protein and polysaccharide concentrations C_2 and C_3 (wt %), respectively, which are represented on the phase diagram by points lining on the secant [C^o_2, C^o_3]. The phase state of the systems was determined by eye after centrifugation (60,000 g for 1h at 298 K) following a rest period of 1h at 298 K. When phase separation occurred, the volumes of the two coexisting liquid phases were determined by weighing, assuming the density of the phases to be equal to 1.0. The points where the binodal intersects the secant [C^o_2, C^o_3] are given by extrapolation of the relation $r = V'/(V' + V'') = f(C_2/C^o_2)$ to $r = 0$ and to $r = 1$; V' and V'' represent volumes of the LBG-enriched (lower phase) and the MC– enriched (upper phase) phases, respectively. The value $r = 0.5$ gives the position of the middle of the tie lines. By repairing the procedure for a series of values of C^o_2 and C^o_3 , the phase diagram can be reconstructed from the set of points corresponding to the binodal and centre of the tie lines. The phase composition of the separated systems was derived from both the material balance of the systems studied and the position of the binodal in the plot, using the correlation:

$$V''/V' = d'\cdot X/d''\cdot Y, \qquad (1)$$

where V'' and V', and d'' and d' - the volumes and density of the polysaccharide enriched and casein enriched phases respectively; X and Y are the distances in the plot between the points corresponding to (i) the composition of protein enriched phase and the composition of initial mixture (X), and (ii) the composition of polysaccharide enriched phase and the composition of the initial mixture (Y). The phase composition of the separated systems was checked on the basis of the system's material balance.

The calculation of the phase compositions and the plot of phase diagrams were performed using the Origin 5.0 (Microcal, USA) software. The error of determination of the critical point and phase composition by this method did not exceed 5-7%.

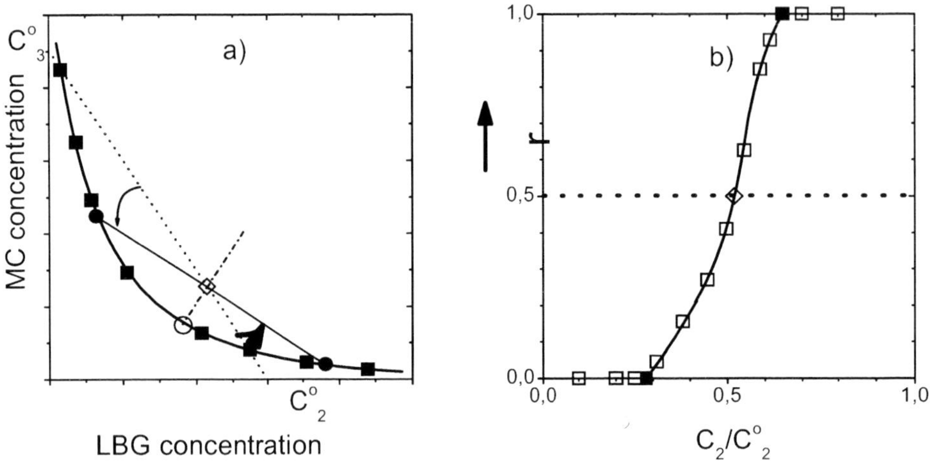

Figure 3. Illustration of the phase volume ratio method

The compatibility of caseins and guar gum was characterised by the co-ordinates of the threshold (C_2^*; C_3^*; $C^*_t = C_2^* + C_3^*$), and critical ($C_{2,c}$; $C_{3,c}$; $C_{c,t} = C_{2,c} + C_{3,c}$) points, as well as by the values of the maximum solubility of polysaccharide in a concentrated solution of protein - $\bar{C}_3$ - and the maximum solubility of protein in a concentrated solution of polysaccharide - $\bar{C}_2$ /limit of solubility. The threshold point was determined from the plot as the point where the line with the slope -1 is tangent to the binodal. The critical point of the system was defined as the point where the binodal intersects the rectilinear diameter, which is the line joining the centre of the tie lines (Figure 3).

RESULTS AND DISCUSSION

The isothermal phase diagram of Water -LBG- MC system is presents in Figure 4 in the Cartesian co-ordinates. The main parameters of the diagram obtained are presented in Table 1.

Table 1. Parameters of the phase diagram water-LBG-MC system

Phase composition, wt%				Coordinate of the critical point	
LBG-enriched phase		MC –enriched phase			
LBG	MC	LBG	MC	$C_{2,c}$	$C_{3,c}$
1.58	0.01	0.020	1.56	0.24	0.38
1.23	0.02	0.019	1.46		
0.88	0.014	0.020	1.37		
0.57	0.016	0.06	0.86		

Table 1. Parameters of the phase diagram water-LBG-MC system (Continued)

$K_s = C_{2,c} / C_{3,c}$	Threshold concentration, (C_2*, C*$_3$) wt%		Limit of solubility, wt%	
	LBG	MC	C'_2	C'_3
0.63	0.39	0.15	0.02	0.01

Low values of total threshold concentration ($C^*_t = 0.54$ wt%), limit of solubility of LBG in concentrated solution of caseinate ($\bar{C}_2 = 0.02$ wt%) and limit of solubility of MC in concentrated solution of LBG ($\bar{C}_3 = 0.01$ wt%) allow total concentration of biopolymers in the critical point ($C_{c,\ t} = 0.62$ wt%) and insignificant asymmetry (K_s =0.63) are the peculiarities of the phase diagram obtained. This gives indication on a very low compatibility of these biopolymers and on insignificant differences between the second virial coefficients, characterising LBG - water and MC - water interactions.

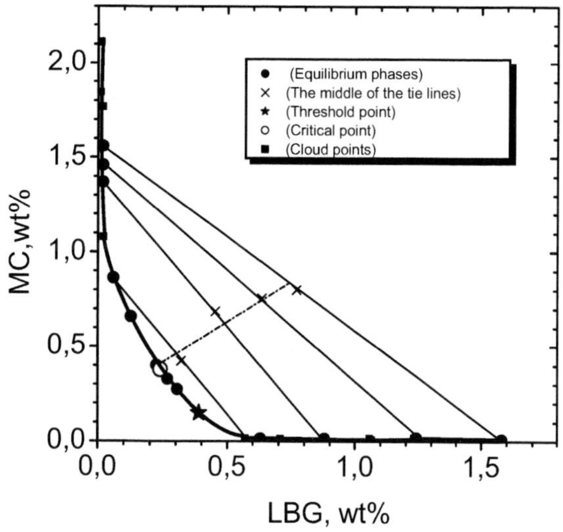

Figure 4. Isothermal Phase diagram of the water-MC-LBG system: pH 5.0; 293 K

References

[1] Tolstoguzov VB In: Hill SE, Ledward DA, and Mitchel RJR. Editors. *Functional properties of food macromolecules.* Gaitherburg Maryland: Aspen Publishers Inc, 1998.

[2] Yurij Antonov, Jacques Lefebvre and J-L. Doublier (1999) *Journal of Applied Polymer Science,* Vol. 71, 471.

[3] Koningsveld, R.; Staverman, A. J. *J Polym Sci,* Part A-2 1968, 6, 305.

[4] Polyakov, V. L; Grinberg, V. Ya.; Tolstoguzov, V. B. *Polym Bull* 1980, 2, 157.

[5] Whistler, R. L. *Industrial Gums*, 2 nd ed.; Academic Press: New York. 1973.

In: New Trends in Biochemical Physics Research
Editors: S. D. Varfolomeev et al., pp. 37-50
ISBN 1-60021-463-0

Chapter 5

APPLYING QUANTUM-CHEMICAL METHODS IN STUDYING STRUCTURE AND FUNCTIONS OF THE PHOTOSYNTHETIC OXYGEN EVOLVING COMPLEX

A. S. Rychkova
Department of Biophysics, Faculty of Physics,
M. V. Lomonosov Moscow State University, Moscow, Russia

ABSTRACT

A detailed quantum-mechanical analysis of a model of a water-oxidizing complex, based on recent X-ray data on the photosystem II structure, is presented. For the first time, a mechanism of water oxidation is proposed and explained within the framework of this model. The role of three manganese atoms, which are not directly involved in water oxidation; the role of cubic structure of the complex; and the necessity of Ca and Cl ions for water oxidation are discussed. Theoretical computations of energies of the complex were made for each of the *S* states.

Keywords: oxygen formation, photosynthesis, water-oxidizing complex, ligand field theory

INTRODUCTION

Photosynthesis, the synthesis of organic compounds from carbon dioxide and water, is driven by the energy of solar radiation. As a byproduct of photosynthesis, molecular oxygen is formed and evolved into the atmosphere. The origin of photosynthesis dates back two to three billions of years; in that time, it has become the most important process on Earth, because it supports all heterotrophic organisms, including humans. The water-oxidizing complex ensures the following reaction: $2H_2O = O_2 + 4H + 4e^-$ [1]. To drive this reaction, four light quanta are successively absorbed. The absorption of a light quantum by the reaction center is accompanied by charge separation: reduction of pheophytin and oxidation of the primary donor P680. The redox potential of an oxidized primary donor (approximately 1.2 V) is extraordinarily high for biological substances and is sufficient for water oxidation. It oxidizes the redox-active tyrosine-161 residue Y_Z of the integral polypeptide D1 (33 kDa). The polypeptides D1, D2 (33 kDa), and others constitute the reaction center of photosystem II. The oxidized electron carrier Y_Z oxidizes the manganese cluster. Successive absorption of four light quanta is accompanied by consecutive removal of four electrons from the manganese cluster, which results in oxidation of two water molecules by the manganese cluster and its reduction to the initial redox state, thus completing the catalytic cycle. This catalytic cycle is called the *S* cycle. It consists of the following five *S* states: S_0 to S_4. Each *S* state corresponds to a certain oxidation state of the cations constituting the manganese cluster.

What is the problem of oxygen formation in photosynthesis? It is not clear how it is possible to remove four protons and four electrons from two water molecules upon successive absorption of four light quanta and, while avoiding the formation of reactive oxygen species ($^{\bullet}O_2^-$, H_2O_2, $^{\bullet}OH$), produce a stable molecule of oxygen. The water-oxidizing complex (WOC) is so efficient in its function that it does not need help from additional defense systems such as antioxidants, superoxide dismutases, and catalases.

For a long time, the only information available on the structure of the WOC was that it includes four manganese ions, one calcium ion (Ca^{2+}), one or more chlorine ions (Cl^-), and redox-active tyrosine Y_z [1]. At the end of January 2004, the Protein Data Bank published new X-ray data on photosystem II with a 3.5-A resolution [2]. The authors suggested a sufficiently detailed structure of the Mn cluster and its environment (Figure 1). The structure of photosystem II was already available from two sources: the structure with a 3.8-A resolution was published in 2000 [3], and that with a 3.7-A resolution became available in 2002 [4]. However, the authors of these structures were unable to describe the WOC structure in detail. The only thing that could be distinguished was the position of the manganese ions and their immediate neighbors, the amino acid residues D170, E33, and A344. In the structure published in 2002 [4], even the Ca^{2+} ion, known [5] to play the key role in water oxidation, was not seen. The latest structure, published in 2004 [2], was very successful. It allowed for the location of four Mn ions, one Ca ion, and the surrounding amino acids. On the basis of the distances between the metal ions and their coordination properties, the authors of [2] suggested that the WOC includes a cubic cluster of Mn_3CaO_4 (Figure 1).

Various models of the WOC and schemes of redox transformations during transitions between the *S* states in these models were suggested from the following experimental data [5–8]: the kinetics of oxygen evolution in response to short light flashes, optical absorption spectra of the WOC intermediate states, electronic paramagnetic resonance spectra of paramagnetic states of the WOC, the fine structure from extended X-ray absorption spectroscopy, and X-ray spectroscopy data on the WOC structure. However, the suggested models are in conflict with the data on the WOC structure described in the recent paper [2].

Hence, in this paper we study the WOC model built on the recent X-ray data. The goals of this study were the following: to analyze the WOC structure using the quantum-mechanical approach; to propose a mechanism of water oxidation for the recently published structure of the WOC, since this was not done in the previous works; and to verify the suggested mechanism by calculating the system energy at each stage of the *S* cycle.

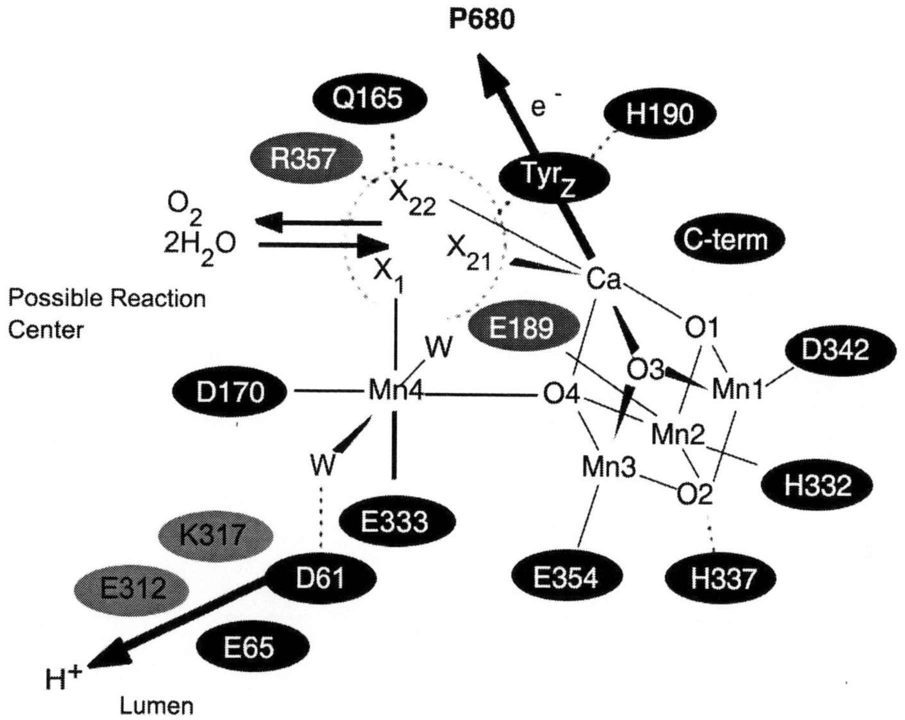

Figure 1. Schematic drawing of the WOC built from X-ray data. X_{11}, X_{22}, X_{21} are hypothetical positions of water molecules, which were obtained from the location of a nonprotein ligand and the coordination model for Mn and Ca. W designates hypothetical water molecules that could not be located at this resolution.

EXPERIMENTAL

The studied system is a coordination compound, i.e., a compound with donor–acceptor bonds. In this compound, one of the atoms (the acceptor) has an incomplete electron shell, whereas the other atom (the donor) has a lone electron pair that fills a vacant shell of the acceptor. To describe such systems, the following two basic theories are available: the crystal field theory and the ligand field theory [9, 10]. The crystal field theory approximates an inorganic complex by a stable multiatom system in which ligands have a purely electrostatic influence on the central ion. The ligand field theory, which takes into account the effect of covalence, is a generalization of the crystal field theory. This method is essentially the same as the well-known method of molecular orbitals (MO), which is widely used in quantum chemistry. The ligand field theory allows the energies of molecular orbitals to be calculated and gives the types of wave functions for coordination systems. We used this theory to analyze the WOC structure.

To calculate the system energy, we used the HyperChem 7 software package and the semiempirical method ZINDO (Dr. Zenner's method of intermediate neglect of differential overlap) [11, 12], since this was the only method in this software package that was parameterized for transitional metals.

RESULTS AND DISCUSSION

Quantum-Mechanical Analysis of the WOC Model

Let us analyze the WOC structure using the ligand field theory [10]. According to the structure described in [2], the Ca ion and one of the Mn ions have five neighbors; two other Mn ions have four neighbors each; and the fourth Mn ion has six neighbors. However, these data could be incomplete (3.5A resolution), and the structure presented does not resolve the Cl ion, which is known from experiments to play a significant role in the operation of the WOC [13]. Additionally, water and hydroxyl molecules could not be located at 3.5A resolution, but they might be essential for WOC functioning. Thus, metal ions have four to six neighbors, which corresponds to the following configurations: tetrahedral, square pyramidal, and octahedral. We will concentrate on the arrangement of molecular orbitals in the octahedral complex (Figure 2) [10], because the relative positions of the energy levels in the other two configurations will be the same, although the energy value at each level will shift. Figure 2 shows that the following stable levels are formed after combining the energy levels of the central atom and the ligands and filling them with electrons: the nondegenerate a_{1g} symmetry level, the triply degenerate levels t_{1u} and t_{2g}, and the doubly degenerate level e_g. Thus, there are nine levels in total, in which 18 electrons could be placed. Additional electrons will occupy antibonding orbitals, which will destabilize the system. Therefore, for transitional metals the eighteen-electron rule has been formulated: in stable complexes of transitional metals ML_n, the total number of electrons in the bonding orbitals (M–L bonds) and nonbonding electrons of the metal is 18. Unlike manganese, calcium is not a transitional metal, because its 3*d* level is not completed, while its 3*s* and 3*p* levels are fully completed. Therefore, we will not consider the influence of the 4*s* and 4*p* orbitals of calcium on the overall distribution of energy levels. In this case, we obtain the following bonding molecular orbitals: the doubly degenerate e_g orbital and the triply degenerate t_{2g} orbital; i.e., there are five orbitals in total, in which ten electrons can be placed. Thus, for calcium we have the ten electron rule instead of the eighteenelectron rule.

Let us apply these two rules to our system. Figure 3 shows the WOC in the S_0 state. Consider that the charge of Mn1, Mn2, and Mn3 is 3+, whereas the charge of Mn4 and Ca is 2+. This distribution of charges agree well with the experimental data on the presence of only one Mn(II) in the S_0 state [14]. However, the idea that the other three manganese ions are in the oxidized state Mn(IV) is widespread in the literature [8]. It is noteworthy that the Mn cluster with its ligands should be totally neutral because of the low permittivity of proteins [15]. Therefore, the environment should be chosen so that the total electric charge of the system is zero. If we assume the charges of the three manganese atoms to be +4, the neutralization of the system will be very problematic. This reasoning makes an obvious case for the charge of three of the manganese atoms to be +3. Note that each oxygen in the corners of the cube ($O1^{2-}$, $O2^{2-}$, $O3^{2-}$, $O4^{2-}$) can share its electron pair with its neighbors, because it has six electrons on the 2*p* level.

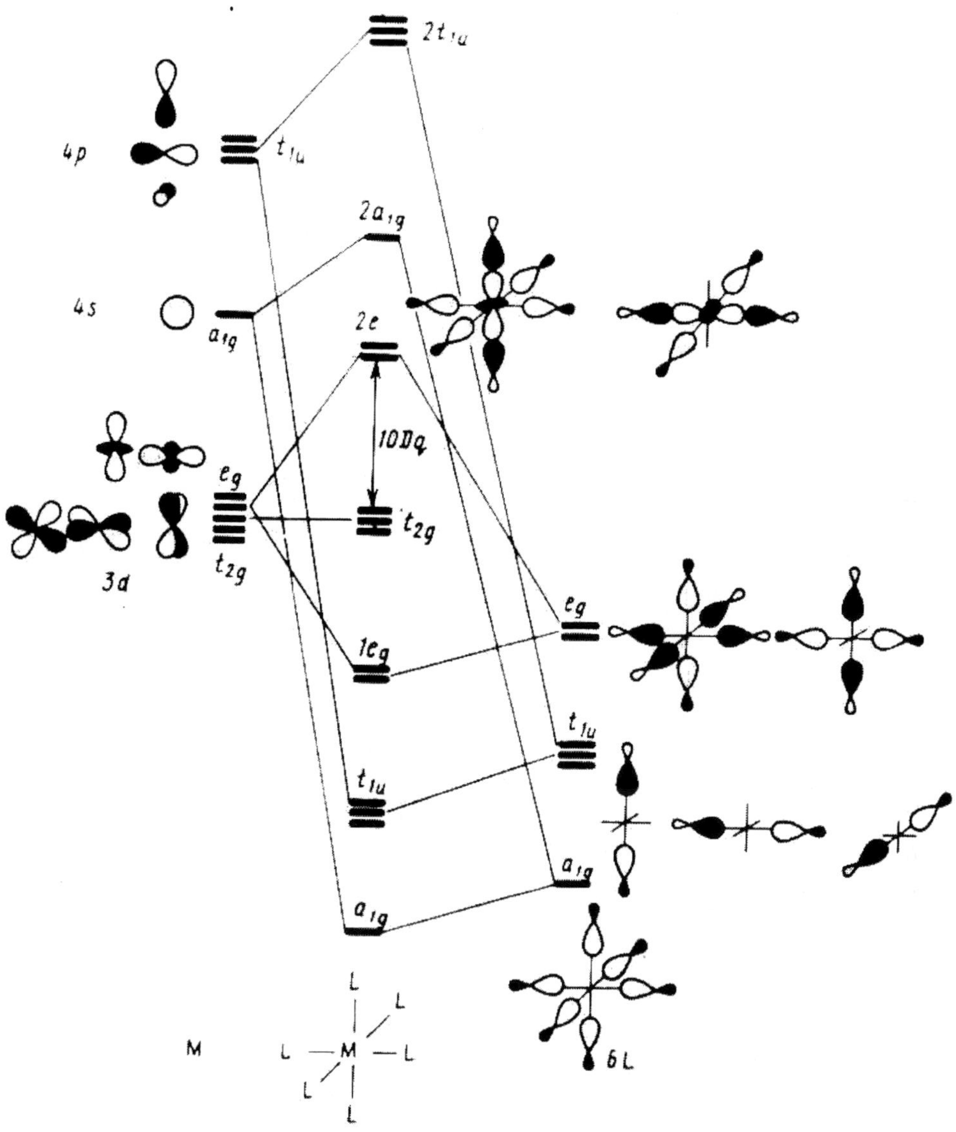

Figure 2. Correlation diagram of MOs in the octahedral complex ML_6^{k+}, built in valent orbitals of metal and group orbitals of ligands. Group orbitals of ligands with octahedral symmetry are drawn separately.

Let us start with the calcium ion. It has five neighbors, from which it can obtain two electrons to get ten in total—the ten-electron rule holds true. Then, we continue with Mn2. The five-element ring of histidine gives five electrons to the central atom; the carboxyl group of glutamic acid gives three electrons; O1, O2, and O3 give six electrons; and the four *d* electrons of Mn2 ion give exactly 18 electrons in total. In other words, the eighteen-electron rule holds true. Consider the Mn1 ion. The situation with this ion is more complicated, because it has no histidine ligand with five electrons. In total, the Mn1 ion gets nine electrons from all its neighbors, which totals 13 electrons, counting its own.

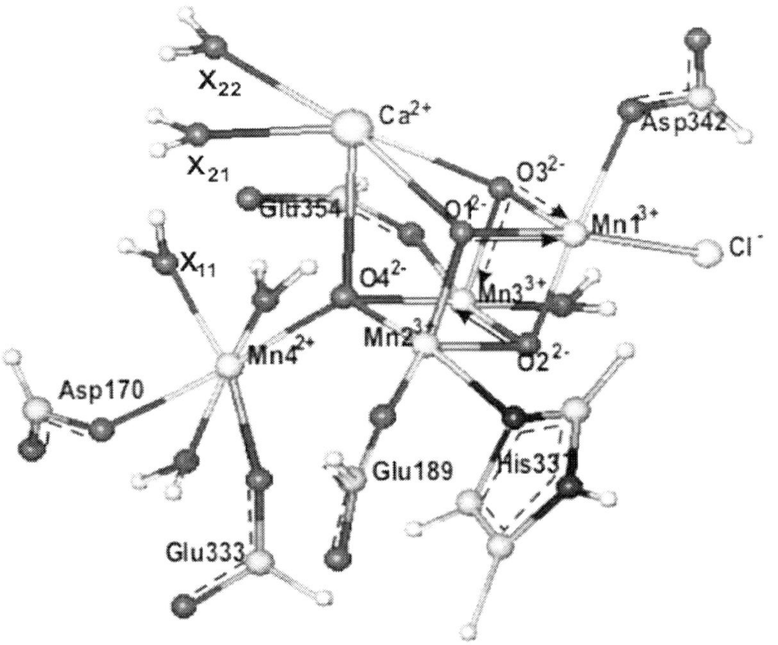

Figure 3. Model of WOC built from the X-ray spectroscopy data. The WOC is shown in the S_0 state.

Where can five more electrons come from? Here we can appeal to the Cl^- ion, which was not located in the X-ray structure. Besides the lone electron pair, it brings into the system the additional negative charge that was lacking for a complete neutralization of the system, since we had 14 positive but only 13 negative charges. Now the total charge of the system is zero, but there are still not enough electrons for Mn1. Let us take a closer look at the three neighboring oxygen atoms O1, O2, and O3. We assume that each of the oxygens shares a pair of *p* electrons with its neighbors, but we should not forget that, along with six 2 *p* electrons, it also has a pair of 2 *s* electrons. After getting an additional pair of electrons from O1, Mn1 lacks only one electron to complete the filling of its orbitals. Keeping this result in mind, consider the Mn3 ion. Like Mn1, it lacks five electrons. But now we cannot use Cl^-, since we no longer need an additional negative charge. However, we can assume that there is a water molecule in close proximity, and it can share its lone electron pair without changing the total charge of the system. Another electron pair could be obtained from O2. Then, as in the case of Mn1, we have 17 electrons. The O4 ion could not help us because it gives its fourth lone pair to the Mn4 ion. There are a few more ions, namely, O3, with the last available lone pair, and the Mn1 and Mn3 ions, which both lack a single electron. What do they do? Certainly, they disrupt the bond between the electrons of oxygen and thus obtain one electron. Finally, the building-up rules for Ca and three Mn ions hold perfectly. We now proceed to the last Mn4 ion. Figure 3 shows that the number of electrons in the molecular orbitals of the complex with the central Mn4 ion is 19; i.e., one electron has to occupy the antibonding orbital, which will destabilize the system. A bond with one water molecule could be disrupted, but in this case there would be 17 electrons in the molecular orbitals of the complex and one electron would be lacking for complete filling of the orbitals. Thus, the

WOC in the described *S*0 state is not stable. To increase the stability of the system, one electron should be either removed or added.

A New Hypothesis of Photosynthetic Oxygen Evolution

Let us now see what will happen upon subsequent removal of four electrons and protons from the system, i.e., consider the whole S_i cycle of the WOC states, which is shown in Figure 4.

$S_0 \rightarrow S_1$

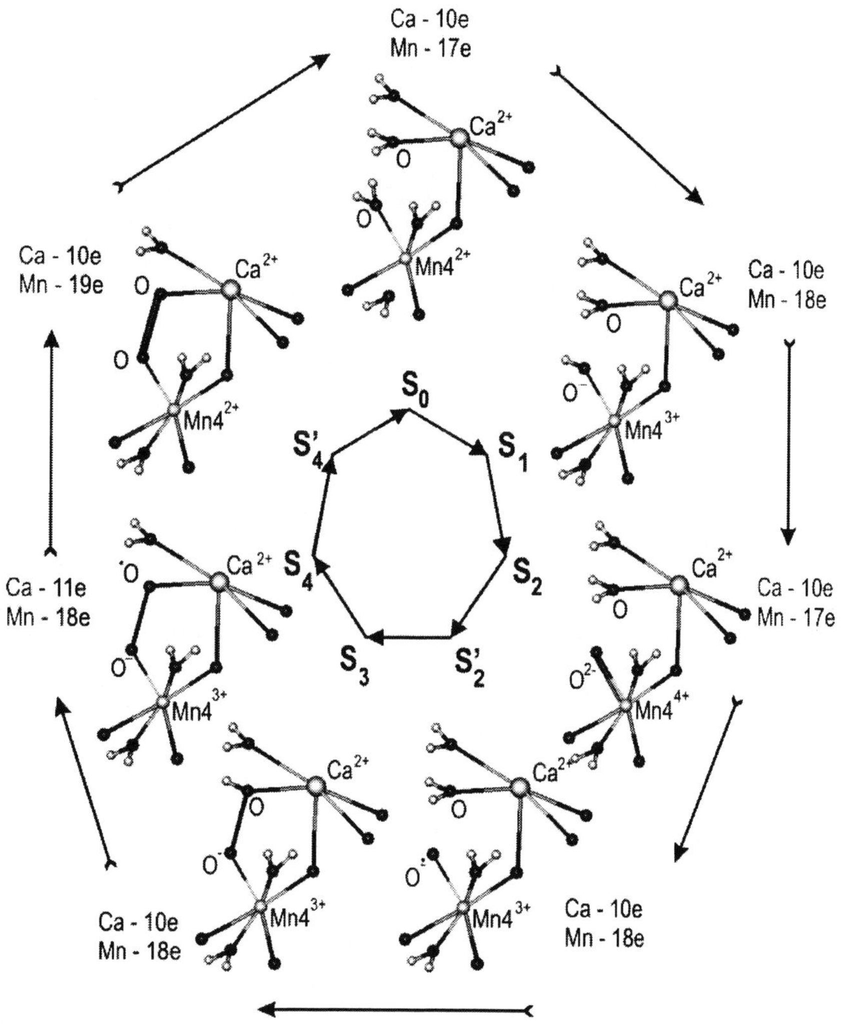

Figure 4. New hypothesis of photosynthetic oxygen evolution. The successive S_i states are shown. Mn4 and Ca are the only ions of the complex shown in the figure, because the other manganese ions do not participate in the reaction and do not change their oxidation states. The number of electrons occupying each MO of Ca and Mn4 are shown for each state.

The S_0 state has already been considered above. We found that all the bonding orbitals of the complex are filled, except the t_{2g} orbital of Mn4, where either one electron is missing, if the bond with water is disrupted, or there is an extra electron, if the bond is not disrupted. After excitation of P680, an electron is transferred from the pigment to the primary acceptor pheophytin. At the same time, the reaction center is oxidized. For its reduction, takes an electron from the nearby tyrosine Y_Z. Then, tyrosine gives H^+ to histidine H190 and becomes a radical $Y_z^{\bullet}$. Thereafter, the electron and the proton leave the Mn cluster, and tyrosine returns into the Y_z state. Thus, after all these transitions, one electron and one proton are missing in the cluster. From where can they be most easily obtained? If we turn back to Figure 1, we see that tyrosine Y_z is located at the side of Ca2+ and is bound to its water X21 by a hydrogen bond. Therefore, a proton and an electron will most likely be removed from X21. Oxygen will obtain an unpaired electron, which will tend to pair up with calcium electrons. However, calcium is very stable, since its shells have exactly ten electrons. Therefore, it is unfavorable for calcium to give one electron to oxygen. The following solution can be suggested: the X21 water molecule can attract a hydrogen from the X11 water molecule. Then, calcium would still have ten electrons in its MO, but there would be changes in the MO of the Mn4 ion. Since hydrogen was pulled away from the X11 water, the oxygen atom obtains an unpaired electron. The Mn4 ion can donate its electron to oxygen and change its oxidation state to Mn(III) with exactly 18 electrons in its MO. Thus, all of the MOs of the complex are completely filled in the *S*1 state, which means that this state is the most stable. This conclusion agrees well with the available experimental data indicating that long-term dark adaptation turns all reaction centers into the *S*1 state [16].

$S_1 \rightarrow S_2$

Let us consider the following transition. The second light quantum reaches the reaction center and excites P680. Following the reasoning presented above, we get the following picture: the X11 water molecule has both hydrogens removed; i.e., there is only one oxygen, and one electron again comes to the oxygen from Mn4. The oxidation state of manganese became Mn(IV), and it has 17 electrons in its MO. It is possible that Mn4 will not give one electron to oxygen and thus will have a completely filled MO, but then oxygen will have an unpaired electron. This situation is shown in Figure 4 as S_2'. Thus, Mn4 changes its oxidation state from Mn(III) to Mn(IV) during the transition from *S*1 to *S*2. At the *S*2 stage, there are two oxidation states, Mn(III) and Mn3(IV), which is in agreement with experimental data [17].

$S_2 \rightarrow S_3$

Upon the following transition, one proton is again removed from the X21 water molecule, but, since X11 does not have any hydrogen atoms, an unpaired electron remains on the oxygen of the X21 water molecule. The oxygen of X11 also has an unpaired electron. These two electrons can become paired, thus forming the initial bond between the oxygen atoms. Again we obtained a very stable state: the Mn4 ion has 18 electrons, and the Ca ion has 10 electrons. It is noteworthy that, according to the X-ray data, the distance between the O11 and O21 oxygen molecules is 2.35 A; whereas in the dioxygen molecule this distance is 1.14 A and in hydrogen peroxide it is 1.24 A. In other words, oxygen atoms should draw closer to each other during the formation of the O–O bond. In principle, this could be achieved by a slight shift of Ca and a slightly higher displacement of the Mn4 ion, which is not so tightly fixed. Such displacements could slightly distort the cubic structure, which is presumably the reason why the short distances between manganese ions increase during the transition from *S*2 to *S*3 [18].

$S_3 \rightarrow S_4 \rightarrow S_0$

The last hydrogen is removed from X21. An unpaired electron reappears at the oxygen atom and pairs with the electron of the O11 oxygen. As a result, a tentative dioxygen molecule is formed and replaced by two water molecules, and the complex reverts to the *S*0 state.

Which experimental results can be explained by our quantum-mechanical approach? It is known that only the even states *S*0, *S*2, and *S*4 give electron paramagnetic resonance signals [19]. Indeed, Figure 4 shows that unpaired electrons are available only in these states, whereas in the other states the MOs of Mn4 and Ca ions are completely filled. There are data showing that calcium is a necessary component of the WOC, because in its absence the only state that can be reached is *S*2*Yz* [20]. Indeed, if at the first two stages hydrogen atoms were removed directly from the X11 molecule, *S*2*Yz* could be reached, but no further transition is possible. It is also known that the Cl– ion could be removed from the sample without inhibition of WOC activity. This means that the Cl– ion is not as crucial an element of the WOC as are the Mn and Ca ions. Let us see what happens if we eliminate the Cl– ion from our model. First, the number of electrons in the MO of the Mn1 ion will change, which will decrease the stability of the system. This problem can be solved by binding a water molecule, as in the case of the Mn3 ion. Second, the total electrical charge of the system will change; i.e., the system will no longer be neutral. But if we assume that a change in the total charge of the complex by one unit will have no significant effect on water oxidation, then the chlorine ion could be neglected. Thus, the suggested hypothesis of water oxidation is in agreement with the majority of the known experimental data on the WOC.

Role of the Three Manganese Ions not Directly Involved in Water Oxidation

To understand the role of the three remaining Mn atoms, one should recall our previous reasoning. We said that the oxygen atoms located at the corners of the cube give electron pairs to Mn1, Mn2, Mn3, and Ca to fill the MOs of their coordination systems. The ions O1, O2, and O3 give four electron pairs to the Mn and Ca ions inside the cube, while O4 has one electron pair and gives it to Mn4. It is important that the O4 ion has *only one* electron pair for Mn4, and it has no single electron. This means that, if here will be a missing electron in the MOs of the Mn4 ion, the latter will not be able to fill its MO with the electrons from the nearby O4. What would happen if oxygen had an additional electron that could be transferred to thc Mn4 ion? In this case, manganese could accept this single electron in the *S*0 state, when the MOs of Mn4 are occupied by 17 electrons. Then, the complex would be in the most stable state, and it would be very difficult to remove an electron and a proton from it. This means that the addition of a single electron to any of metal ligands would break the complex function and prevent water oxidation.

Role of the Ca ion in Water Oxidation

Let us consider why Ca is used for water oxidation. To answer this question, we should remember that Ca is not a transitional metal: it precedes them in the row of chemical elements. Its electron structure is *3s23p64s2*, and its *3p* and *4s* orbitals are completely filled, whereas its *3d* shell is empty. Unlike the Mn ion, the Ca ion does not easily change its oxidation state. This explains the higher affinity of O11 oxygen for an unpaired electron than the affinity of O21 oxygen, as well as why, at each stage after the removal of hydrogen, the X21 water molecule removes hydrogen from the X11 water molecule.

Consider the *S*4 state of the WOC. If the calcium ion did not have a low affinity for an additional electron, the Ca ion would attract an unpaired electron from oxygen to its MO. Then, the *S*4 state could not proceed to the state and oxygen would not be produced, meaning that the second bond between oxygens would not arise. In our view, this explains the role of the Ca ion in the water oxidation system.

Role of the Cubic Structure of the WOC

Presumably, the cubic structure of the WOC is used both to prevent the three manganese ions constituting the cube from changing their oxidation state and to eliminate the influence of this change on water oxidation, because, as was mentioned above, no additional electron can be introduced into the system. These three manganese atoms are incorporated into the cubic structure and, due to their environment, have exactly 18 electrons in their MO, which prevent them from influencing water oxidation. Presumably, the cubic structure of the cluster is also the proper geometry for WOC function. The location of the Mn4 ion at some distance from the cubic cluster is also nonrandom: it facilitates the mutual approach of Mn4 and Ca ions at the stage of bond formation between the O11 and O21 oxygens.

Estimate of the System Energies in the *S* States

It is important to verify the proposed hypothesis of water oxidation by theoretical computations. For this purpose, we used the software package HyperChem and the semiempirical method ZINDO, because this was the only method applicable to systems with transitional metals.

To obtain the system energy, it was necessary first to optimize its geometry, i.e., to find the configuration corresponding to the minimum energy. We made numerous attempts to optimize the system using various optimization methods, but they all failed. Optimization considerably distorted the structure of the complex, even when the parts of the complex with known locations [2] were fixed. Therefore, we came to the conclusion that the ZINDO method is not suitable for optimization of the geometry of complexes with transitional metals. Although this software did not allow us to optimize the geometry, we calculated the energy of the system at a particular point (i.e., the energy of the particular presented structure). Such a calculation was made for the structure shown in Figure 5. To simplify the calculations, we used HCOO– groups instead of aspartic and glutamic acids, and the group instead of histidine. The results of the computations are presented in the table. The table shows that, although the total energy of the system is negative, the energy of the bond at each stage, except *S*4, is positive. This means that the cluster could not be in a stable state without external stabilization from the protein globule.

Table. Calculated energies of the system in various states of the WOC *S* cycle

	1		2	
	E_{tot}, ckal/mol	E_b, ckal/mol	E_{tot}, ckal/mol	E_b, ckal/mol
S_0	-242300	39672	-293715	-6610
S_1	-202391	90352	-293244	-6440
S_2	-218301	74141	-292578	-6075
S_3	-246060	46081	-292209	-6007
S_4	-292066	-226	-291550	-5649

E_{tot} designates the total energy of the system; *E*b, the bond energy; *1*, the initial distances used in computations, which are shown in Figure 5; *2*, the distances used in computations are 2.0 A.

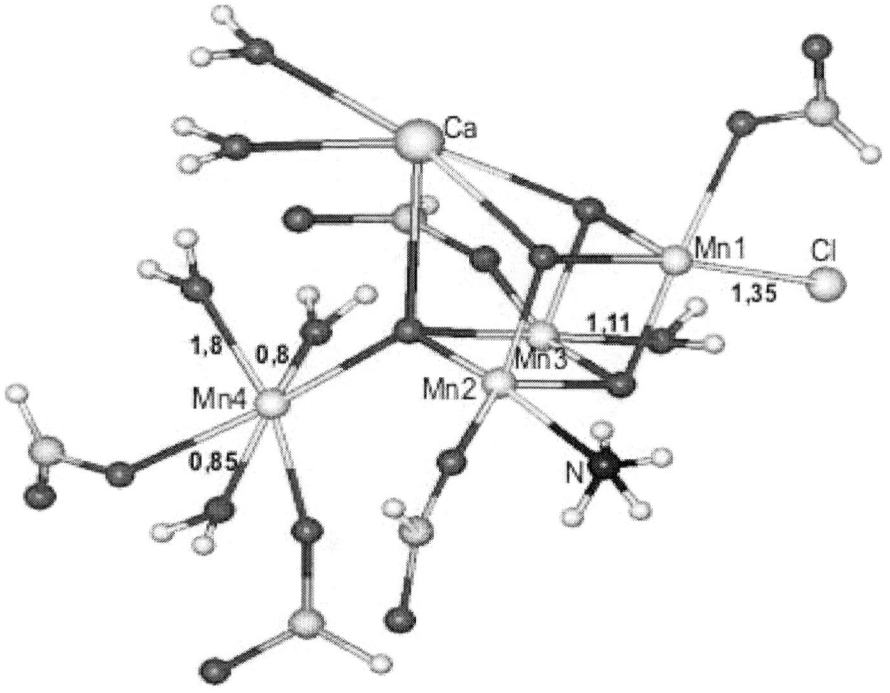

Figure 5. Model of the water-oxidizing complex used in computations of the *S*-state energies. The distances between the atoms are in angstroms.

To obtain steady states of the complex, we tried to optimize the geometry without using any of the available optimization programs. For this purpose, we changed the length of the bonds with those molecules that were not obtained from the structure [2] (four water molecules and the chlorine ion). We assumed that those parts of the complex that were obtained from the X-ray system did not change their position. By varying the distance to these molecules, we obtained negative bond energies for each *S* state. The results are summarized in the table. It can be seen that the bond energy in each of the *S* states is negative. This means that the system with this configuration is stable and can perform water oxidation.

We should note that this is the first time that such a quantum-mechanical analysis of the model based on the recent X-ray data of WOC structure is used in the study of WOC function. Such an approach throws light on the cubic structure of the complex, the role of three manganese ions not directly involved in water oxidation, and the role of calcium in the complex. Application of the eighteen- and ten-electron rules explains most of the available experimental data on the WOC. Taking these rules into account, the model built on the recent X-ray data on photosystem II was extended by the Cl– ion and the water molecule. For the first time, a mechanism of water oxidation was proposed for such a model, theoretical computations of the system energies were performed for each of the *S* states, and a conclusion regarding their stability was drawn.

ACKNOWLEDGMENTS

We are grateful to S.V. Poltev for consultations on the software package and to A.N. Tikhonov for valuable comments on the work. This study was partly supported by the program "Universities of Russia".

REFERENCES

[1] B. Rubin, *Biophysics* (Universitet, Moscow, 2000) [in Russian].
[2] K. N. Ferreira, T. M. Iverson, K. Maghlaoui et al., *Science* 303, 1831–1838 (2004).
[3] A. Zouni, H.-T. Witt, J. Kern, et al., *Nature* 409, 739 (2001).
[4] N. Kamiya and J.-R. Shen, *Proc. Natl. Acad. Sci. USA* 100, 98–103 (2003).
[5] Taka-Aki Ono, *Biochim. Biophys. Acta* 1503, 40–51 (2001).
[6] G. T. Babcock, *Biochim. Biophys. Acta* 1458, 199–219 (2000).
[7] J. H. A. Nuget, A. M. Rich, and C. W. Evans, *Biochim. Biophys. Acta* 1503, 138–146 (2001).
[8] J. S. Vrettos, J. Limburg, and G. W. Brudvig, *Biochim. Biophys. Acta* 1503, 229–245 (2001).
[9] B. Bersuker, *Electron Structure and Properties of Coordination Compounds* (Khimiya, Leningrad, 1986) [in Russian].
[10] V. I. Mishkin, B. Ya. Simkin, and R. M. Minyaev, *Theory of Molecular Structure* (Feniks, Rostov-on-Don, 1997) [in Russian].
[11] M. C. Zerner, *Rev. Comput. Chem.* 2, 313 (1991).
[12] M. C. Zerner, *Theor. Chim. Acta (Berl.)* 53, 21–54 (1979).
[13] G. M. Ananyev, L. Zaltsman, C. Vasko, and G. C. Dismukes, *Biochim. Biophys. Acta* 1503, 52–68 (2001).

[14] H. Wincencjusz, S. I. Allakhverdiev, V. V. Klimov, and H. J. van Gorkom, *Biochim. Biophys. Acta* 1273, 1–3 (1996).
[15] P. E. M. Siegbahn, *Inorg. Chem.* 39, 2923–2935 (2000).
[16] G. Renger, Biochim. Biophys. *Acta* 1503, 210–228 (2001).
[17] G. Renger, *Physiologia Plantarum* 100, 828–841 (1997).
[18] V. L. Pecoraro, M. J. Baldwin, M. T. Caudle, et al., *J. Inorg. Biochem.* 59, 614 (1995).
[19] C. Andrews, R. Cinco, H. Dau, et al., *Physica B: Condensed Matter* 208, 657–659 (1995).
[20] C. Andrews, M. J. Latimer, W. Liang, et al., *J. Inorg. Biochem.* 59, 611 (1995).

In: New Trends in Biochemical Physics Research
Editors: S. D. Varfolomeev et al., pp. 51-63
ISBN 1-60021-463-0

Chapter 6

LOW-K DIELECTRICS - A NEW REALITY IN MICROELECTRONICS: MODERN TRENDS, APPROACHES AND APPLICATION PROBLEMS

***D. Soujon*[a,b], *R. Sharafiev*[a] *and S. Bibikov*[a]**
[a]Institute of Biochemical Physics, Russian Academy of Sciences, Moscow, Russia
[b]Ludwig-Maximilians-Universität München, München, Germany

ABSTRACT

This paper summarizes the introduction of low dielectric constant (low-k) materials in microelectronic devices and points out main application problems.

After a broad overview of new low-k materials we focus on introduction of porosity as key approach to lower down permittivity. Chemical methods to produce porosity by incorporation of porogens, use of solid state solvents and thermal degradation with carbon radicals are discussed.

The main challenge lies in generating low-k dielectrics compatible with modern semiconductor processing. We study two versatile physical methods to characterize mechanical and structural properties: nanoindentation and laser-generated surface acoustic waves.

INTRODUCTION

New Trends

According to Moore's law which held to be true since the 1960s the computing power of chip manufacturing doubles every 18 months. This improvement in performance was achieved by decreasing the nominal feature size and gate width of semiconductor devices. [1]

Driven by cost reduction the aim of the semiconductor industry is to increase the density of circuits by using multilevel metallization structure, increasing the density of transistors and shrinking the lateral dimensions of the Aluminum (Al) / Silicon dioxide (SiO_2) interconnections. [2]

As Aluminum lines approach even smaller dimensions (<0.18 μm in width) the performance of logic integrated circuits is more and more dominated by the interconnect delay caused by the aluminum lines and the SiO_2 insulating material. With the introduction of copper which has ~30% lower resistivity than Al part of the limiting factor has been removed. However the properties of the dielectric material between the layers and the lines have to be addressed. [3]

Therefore industry is actively seeking new interlayer dielectric materials (ILD) with lower permittivity k. The National Technology Roadmap for Semiconductors [4] indicated a need for materials with dielectric constants of 1.5 – 2.0 for dimensions below 0.13 μm by the year 2003. In fact the Intel® Pentium® 4 processors were introduced using a 0.18 μm and 0.13 μm node. Today Intel is testing feature sizes as small as 0.065 μm and predicts processors heading to 1 billion transistors by 2007. [5]

Interconnect Delay

Using a simple approach to model the interconnect delay (or RC delay) the essential role of the dielectric constant of the insulating material becomes obvious.

Figure 1 shows a picture of this model with lateral line to line capacities and vertical layer to layer capacities. The total RC delay can be written as [3]:

$$\tau = RC = 2\rho k \varepsilon_0 L^2 \left(\frac{4}{P^2} + \frac{1}{T^2} \right) \tag{0.1}$$

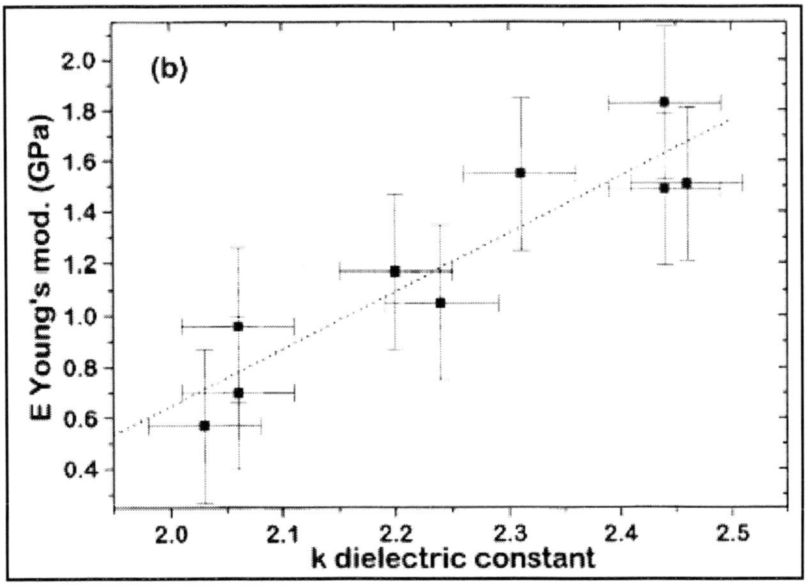

Figure 1. Relationship of dielectric constant and elastic modulus (Young's modulus) [20]

In this equation R is the total resistance, C is the total capacitance, ρ is the resistivity of metal, L is the interconnect line length, P is the line spacing, T is the layer thickness, ε_0 is the permittivity of vacuum and k is the relative dielectric constant.

Equation 1.1 illustrates that conventional thoughts to decrease chip geometry dimensions P and T are forcing the RC delay to increase. Additionally, increasing system complexity drives L up and therefore contributes to RC delay. Therefore a decreasing interconnection delay can only be accomplished by reducing the resistivity of the metal, i.e. replacing Al by Cu for example and addressing the permittivity of the interconnect insulator.

The use of Low-k dielectrics presents also other benefits: reduced power consumption and heat dissipation as well as reduced interline crosstalk. [6]

Overview of New Interlayer Materials

One of the biggest challenges is to find a low-k dielectric material with electrical, thermal and mechanical properties comparable to or better than those of conventional SiO_2. Furthermore, to meet the next generation device requirements a number of integration and reliability issues have to be considered. The most important dielectric requirements include an isotropic dielectric constant, good thermal stability and a low coefficient of thermal expansion. These properties are essential to prevent property changes during subsequent thermal processes. Good adhesion and etch and strip resistance are key factors to withstand Chemical Mechanical Polishing (CMP) during semiconductor processing.

Table 1. Low-k dielectric materials, physical and chemical properties.

Material	Chemical process	Application	Material properties/ Issues	K value
SiO_2	Oxidation of TEOS	CVD	Well known and good properties	3.9 - 4.5
Fluorosilicate-glass (SiOF)	e.g. using Tri-Fluormethylsilan	CVD	Only 10% k reduction	3.2 - 4.0
Polyimides	Imidization of polyamic acids	Spin-On	Good thermal stability Moisture absorption Dielectric constant anisotropy	3.1 - 3.4
B-staged polymers (poly-aromatics)	Resins formed from a prepolymer of DVS-BCB	Spin-On	Low moisture uptake Lack of thermal stability	2.6 - 2.7
Polyarylene ethers (PAE)	Aromatic precursor and Bisphenol	Spin-On	Thermally stable Low moisture uptake Relatively high k	2.6 - 2.8
Inorganic polymers HSQ	SiO_2 backbone with cage or ladder structure	Spin-On	Lack of thermal stability, k increases above 400°C	2.9 - 3.2
Inorganic Polymers MSQ	As HSQ but CH_3 replaces H	Spin-On	Better thermal stability and higher resistance to cracking than HSQ	2.6 - 2.8

Table 1 presents an overview of Low-k dielectric materials with their physical and chemical properties.

The first attempt adopted was to lower the dielectric constant of SiO_2 by doping it with fluorine thus producing fluorinated silica glass (FSG). However to minimize corrosion of the Al lines the fluorine content has to be set below 3% which results in modest reduction of k to about 3.5. [7]

Polyimides are high viscosity engineering polymers that are formed by condensation reaction (imidization) of polyamic acids. They have been widely used as dielectric and packaging materials in the microelectronics industry because of their good mechanical, thermal and dielectric properties.

However with dielectric constants in the range of 3.1-3.5 conventional polyimides are insufficient to meet the requirement for technology nodes below 130nm. ($k< 2.5$) The main issues with polyimides are high moisture absorption and dielectric constant anisotropy. The within plane dielectric constant of fluorinated and non-fluorinated polyimides is significantly higher than the out-of plane dielectric constant. [8]

Other dielectric materials in use are B-staged polymers which are thermoset resins formed from a prepolymer of divinyl siloxane bisbenzocyclobutene (DVS-BCB). BCB polymers have low moisture uptake and low isotropic permittivity k. When they are exposed to air at temperatures of 150-300°C their weight loss exceeds 1% thus they account for insufficient thermal stability for semiconductor processing. (Thermal stability up to 400°C needed). Although fluorination improves thermal stability of BCB, interface issues with metals remain a main impediment to damascene integration of fluorinated BCB materials with copper. [3]

Polyarylene ethers (PAE) are produced from an aromatic precursor (bis-phenol). They are thermally stable with low moisture uptake and have good adhesion to metals. However the dielectric constant is relatively high (around 2.7).

Silsesquioxane are inorganic polymers built on a SiO_2 backbone. They have excellent planarization and gap-filling properties. Two representative silsesquioxane which vary only with a side group, hydrogen silsesquioxane (HSQ) and methyl silsesquioxane (MSQ) have drawn special attention. The molecular structures of HSQ and MSQ are represented in Figure 2. They can both form a ladder or a cage structure.

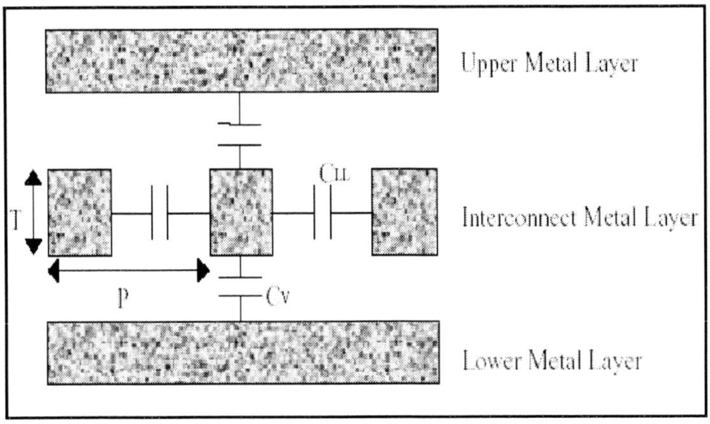

Figure 2. Model of lateral and vertical capacities in multilayer structure [3]

HSQ is a completely anorganic material with Si-H side groups and was first implemented in Dynamic Random Access Memory (DRAM) chips to reduce the cost of processing. The thermal stability of HSQ is one of the main issues as a processing at temperature above 400°C increases the k value near that of SiO_2. MSQ is similar to HSQ but has a 22% organic content resulting from the CH_3 side groups that replace the hydrogen of HSQ. The main advantage of MSQ is that it does not decompose at temperatures above 400°C and is therefore thermally stable. Additionally MSQ has a higher resistance to cracking than HSQ. [3, 9]

Porous Low-k Dielectrics

According to the International Technology Roadmap for Semiconductors (ITRS) [10], device scaling will require to reduce the dielectric constant to about 2.1 for the 45 nm node. This requirement can be achieved by reducing the density of the matrix dielectric incorporating low molecular weight atoms and/or by adding more free volume (air gaps have the lowest k-value at 1) to the structure. [11]

In case of pores with a lower k than the bulk material the Maxwell-Garnett model [12] can be used to determine the fraction of porosity needed to generate the desired k value. The model assumes that the pores are isolated spheres in a matrix. For a matrix k of 4.0 for example 80% porosity is needed to reach an effective k value of 1.5.

This high porosity is not a viable solution as the microscopic pores further degrade the properties of the dielectric materials, particularly the thermo mechanical properties: Mechanical properties (Elastic modulus, hardness and fracture toughness) and thermal conductivity scale inversely with porosity. Therefore a trade-off between low-k and worsening properties has to be found. Another issue is the pore type and the pore size distribution. The pores should be of the close cell type to prevent crack propagation and reduce moisture uptake and the pore size distribution has to be narrow to ensure that the bulk permittivity k is homogenous and isotropic. [3,13]

Table 2 gives an overview of different porous Low-k dielectrics with their respective k value.

Table 2. Porous Low-k dielectric materials and their k values.

Material	k value
Porous Silica (Aerogels/Xerogels)	1.1 - 2.2
Porous HSQ	1.7 - 2.2
Porous MSQ	1.8 - 2.2
Porous PAE	1.8 - 2.2

Aerogels are highly porous solids similar to a gel where the liquid component is replaced with gas. Xerogels are dried out open structures that are formed from a gel.

Ways of Producing Porous Materials

Incorporation of Porogens

One widely used approach to introduce nanometer sized pores into the material relies on the incorporation of an organic supramolecular template (porogen) within a host thermosetting matrix and subsequent heating to high temperatures. This results in thermal decomposition and volatilization of the porogen leaving behind nano scaled pores in the dielectric film.

Star shaped and hyper branched macromolecules that thermally degrade below 400° C are commonly used as porogen. One example of a star-shaped porogen studied by W. Oh et al. [14] is presented in Figure 3.

(a) HSQ

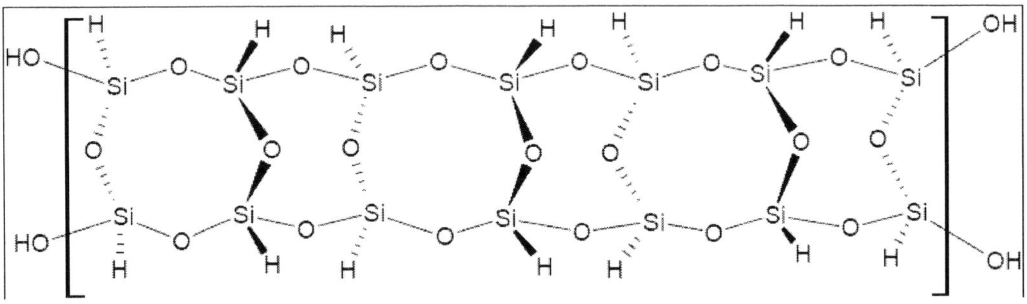

(b) MSQ

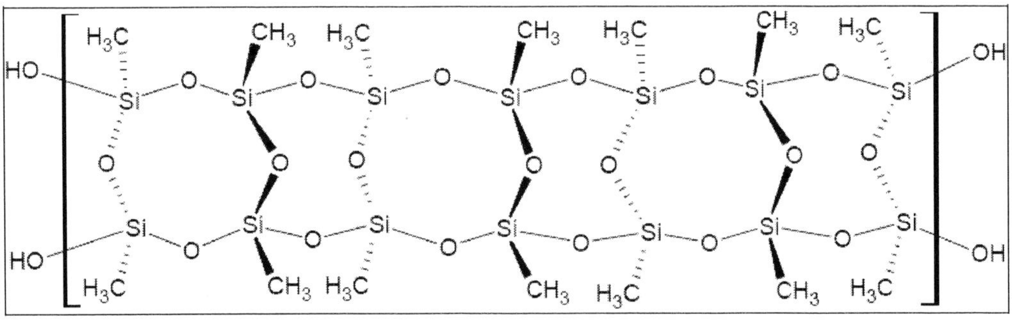

Figure 3. Molecular ladder structure of HSQ (a) and MSQ (b) [3]

Jun Xu et al. [15] have shown that the incorporation of porogen into silsesquioxane matrices allows for easy processability and leads to a homogenous depth profile. The pores size and closeness scales inversely correspond with molecular weight of polymers. Moreover, as illustrated by W. Oh et al. [14] in Figure 4, the refractive index n and the dielectric constant k are linearly reduced with increasing porogen loading in the prepolymer composite.

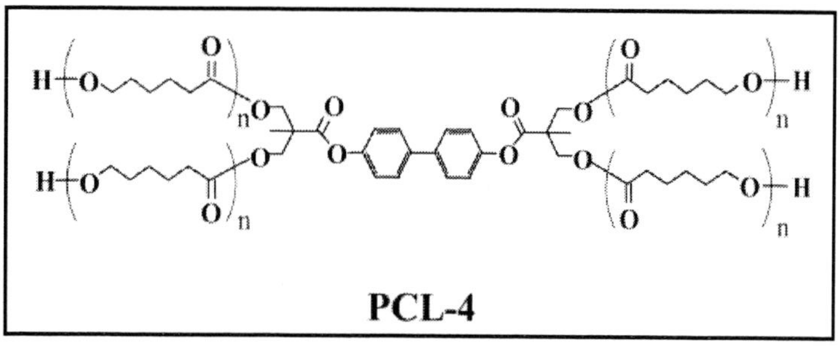

Figure 4. Example of star shaped poly(ε -caprolactone) porogen [14]

HSQ in Solution with Solid-state Solvents

HaeJon Lee et al. [16] illustrate another method to produce porous Low-k dielectric thin films. Films were prepared by spin-coating a solution containing a HSQ resin, a low boiling point solvent (methyl propyl ketone) and a high boiling point solvent (tetradecane) where different compositions of solvents have been examined.

The obtained porous films have a uniform density profile measured by specular X-ray reflection technique (SXR). An increasing mass fraction of high boiling point solvent increases porosity and decreases wall material density. Additionally the average mass density and wall density of the films increase with increasing permittivity k and decreasing Si-H fraction.

Figure 5 shows the X-ray reflectivity curves depending on the Si-H fraction. The density depth profile of the films is determined by a fit to model electron density profiles.

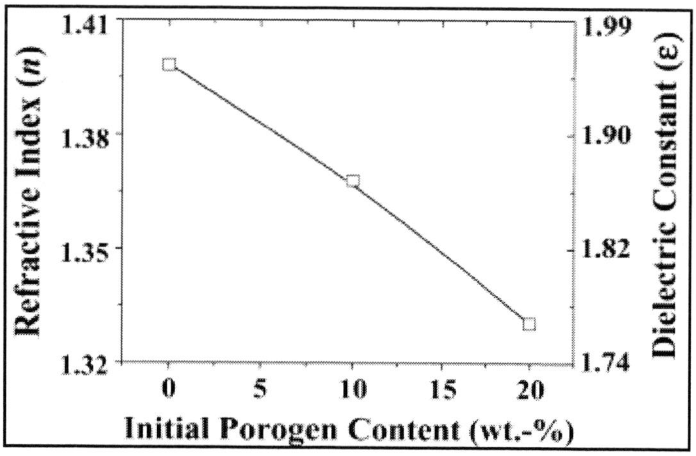

Figure 5. Variation of refractive index n and permittivity k as a function of porogen loading [14].

Thermal Degradation of Siloxane Resin

Duan Li Ou et al have reported a sacrificial approach whereby organic groups are burnt out to generate ultra low-k dielectric films. [17]

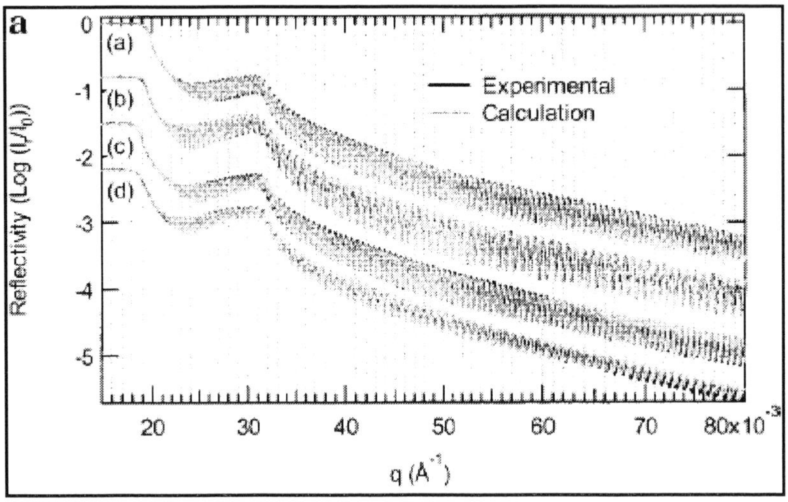

Figure 6. Specular X-ray reflectivity curves of HSQ-based porous thin films with k=2.0 and different Si-H fractions of (a) 30%, (b) 35%, (c) 47%, (d) 52% [16].

The loss of organic groups in form of several vaporized species leads to the formation of void space in the resin matrix during the pyrolysis process. Seven distinguished types of organic functions were linked to siloxane network by covalent bonds. The induced pores were smaller than 50Å and special types of organic functions were needed to create porosity.

The size of the side chain determines the pore size distribution as illustrated in Figure 6.

Physical Properties of Low-k Dielectrics and Methods of Investigation

The main challenge lies in generating low-k dielectric films that are compatible with modern semiconductor processing such as etching, stripping, cleaning, damascence lithography, device packaging processes and chemical mechanical polishing (CMP).

Good thermal stability and low coefficient of expansion is needed to ensure device stability which depends on many factors including the ability of the material to withstand device stresses and the materials adhesion to neighboring structures.

Production of standardized, uniform and reproducible films requires reliable and accessible characterization methods, which allow the determination of critical film properties. The most important factors, determining film properties are density (porosity), elastic modulus, hardness, adhesion and toughness. These structural and mechanical properties can be characterized using nanoindentation and surface acoustic waves (SAW) methods.

Nanoindentation

Nanoindentation is a versatile technique for measuring film's mechanical properties where no special sample preparation is required. Figure 7 demonstrates the principle of the nanoindentation technique: a hard tip, usually a diamond is pressed into the sample with a know load (displacement) which is continuously measured e.g. using a capacitance displacement gage.

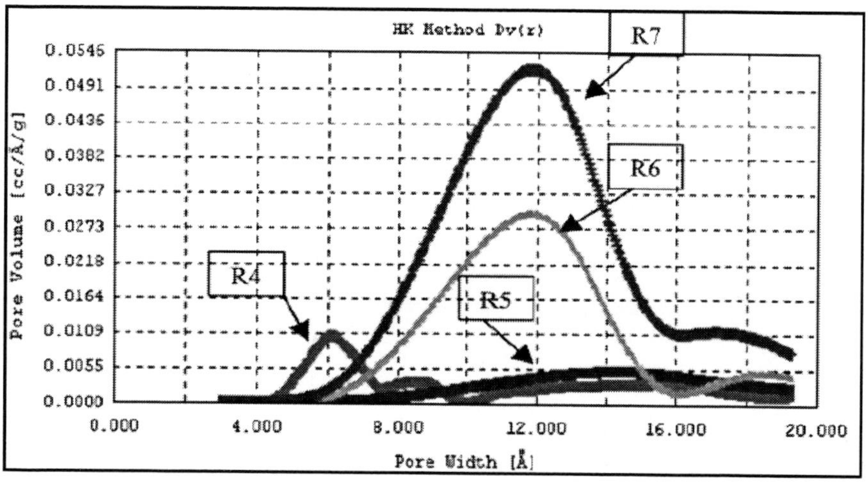

Figure 7. Pore size distributions of selected pyrolysed organically modified siloxane resins [17].

After some time the load is removed and the effect (crack length, blister size, etc.) on the material allows conclusions to be drawn about the mechanical properties. [18]

The elastic modulus and hardness can be directly extracted from the nanoindentation curve that displays the applied load as a function of displacement. The result obtained by Alex A. Volinsky et al [19] for elastic modulus of silica-based porous low-k films can be seen in figure 8. The average value is around 7 GPa and the films have a relatively high hardness of 1.3 GPa. (Not presented in the figure).

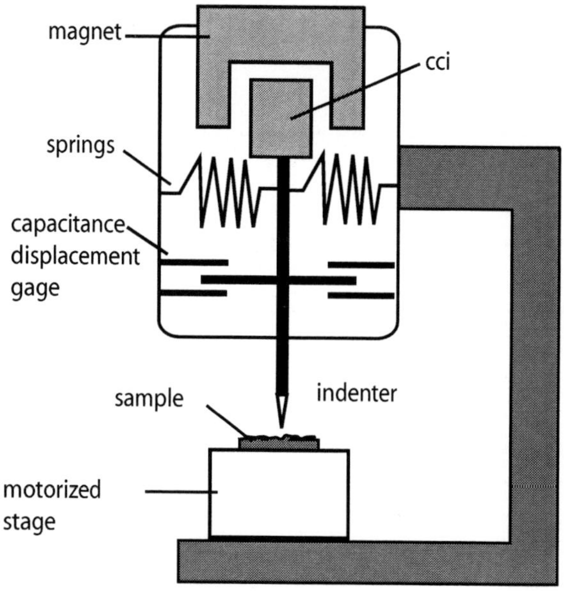

Figure 8. Principle of nanoindentation [18].

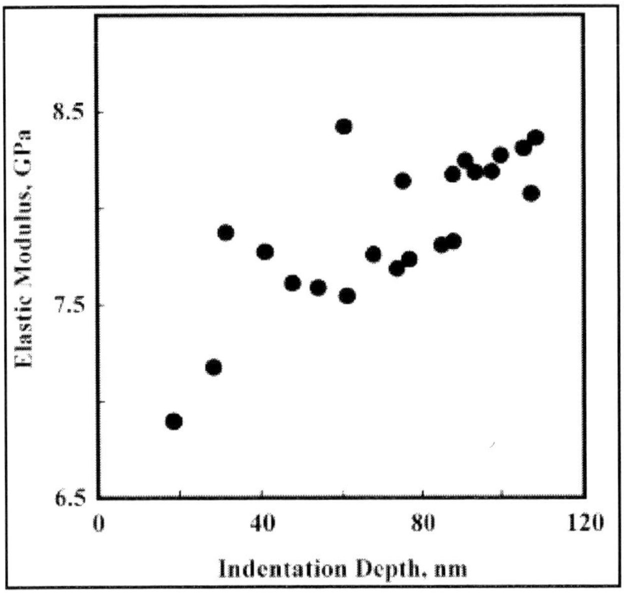

Figure 9. Elastic modulus of a 2 μm low-k dielectric film as a function of indentation depth [19]

Fracture toughness is calculated based on maximum indentation load and crack length with an accuracy of 40%. Figure 9a presents an ideal cube-corner indentation-induced fracture, the maximal measured fracture toughness for the silica-based porous films is as low as 0.05 $MPa \cdot m^{1/2}$. [19]

Finally the adhesion strength, which corresponds to the practical work of adhesion, can be calculated based on the size delamination by high load indentation. In order to reduce the formation of pile-up around the tip, super layer indentation technique is used. A high modulus super layer is deposited on top of the film and the area of the delamination blister (Figure 9b) formed around the indent is used to calculate the practical work of adhesion.

Laser-generated Surface Acoustic Waves

Surface acoustic waves can be used to provide accurate values of porosity and elastic modulus of porous low-k dielectric films. This technique is suitable for process control and allows for non-destructive determination of critical film properties.

In the work of C.M. Flammery et al. [20] a nanosecond laser-pulse is focused into a line shape on the sample surface where the absorbed heat energy generates broadband SAW packages.

The wave packets propagating across the sample surface are detected by piezoelectric foil with a steel-wedge transducer.

As waves of different frequency sample a different proportion of layer and substrate with different elastic properties the wave packets becomes dispersed and the wave velocity is frequency-dependent.

The dispersion curves obtained by Fourier analysis are characteristic of the elastic properties of the layer-substrate system. The shape depends on thickness, porosity and elastic modulus of the film. Dispersion curves for different layer thicknesses are presented in Figure

10. The Layer properties are determined with best fit of one or two parameters to dispersion curve.

Density/ porosity and elastic modulus show a strong correlation with film dielectric constant.

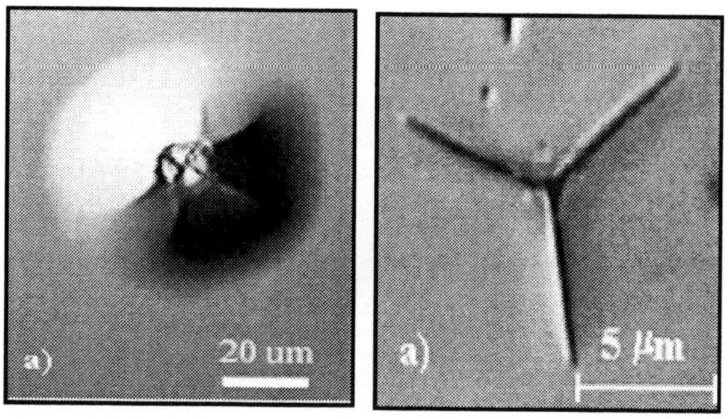

Figure 10. a) Optical micrograph of cube-corner indentation-induced fracture in low-k dielectric films 2 μm thick; b) Nomarski contrast optical image of TiW/400 nm low-k dielectric blister [19]

The elastic modulus is scaling inversely with porosity as illustrated in Figure 11. Aerogel films studied by CM. Flammery et al. had porosities of 60-70% and elastic modulus of approximately 1-2 GPa.

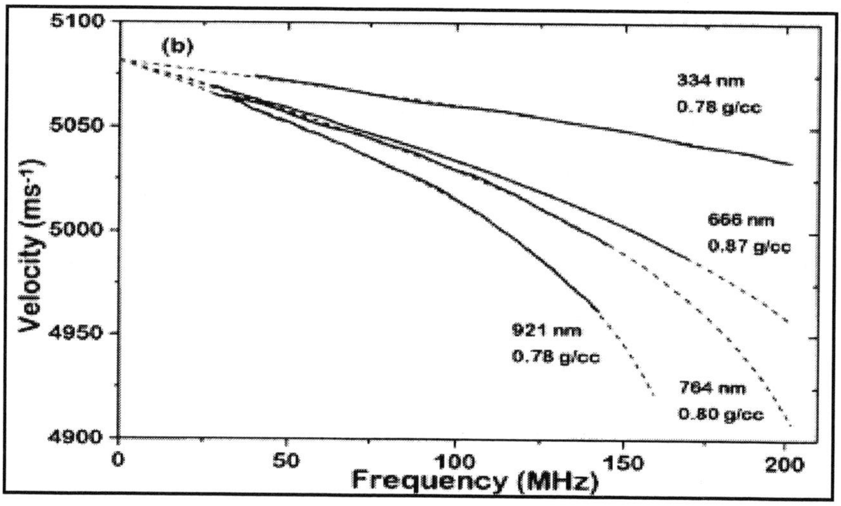

Figure 11. Dispersion curves for different layer thicknesses [20]

Conclusion

As integrated circuit feature sizes continue to shrink new low-k dielectric materials are needed to address problems with interconnect delay (RC delay), power consumption and crosstalk.

One way to produce low-k dielectric films is the introduction of nanometer scale pores to reduce the effective permittivity by incorporation of porogens, combination of solid-state solvents or thermal degradation with carbon radicals.

However the main challenge consists in generating low-k dielectric materials compatible with semiconductor processes such as CMP. Therefore the characterization of mechanical and structural properties is essential.

Nanoindentation and SAW techniques used to investigate porous thin film samples underline that low-k dielectric materials are still far from being ideal for product integration due to poor mechanical properties.

REFERENCES

[1] Bob Gasser, Nanotechnology in Electron Devices, *Intel Corporation electronic publications,* 2003, p. 1-51.

[2] M. Brillouet, Multilevel interconnection technologies and future requirements for logic applications, *Microelectronic Engineering* 37/38, 1997, p. 5-13.

[3] Michael E. Clarke, Introducing Low-k dielectrics into Semiconductor processing, *Mykrolis Application Notes,* 2001, p. 1-20.

[4] The National Technology Roadmap for Semiconductors, L. Wilson, ed., *Semiconductor Industry Association,* San Jose, California, 1997.

[5] Derek McAuley, Where is Moore's law taking us?, *Intel Corporation electronic publications,* 2005, p. 1-25.

[6] C. Wyon, Future technology for advanced MOS devices, *Nuclear Instruments and Methods in Physics Research B* 186, 2002, p. 380-391.

[7] Michael J. Samspon, Technology Readiness Overview - Low-k Interlevel Dielectrics Technology - Brief description of low-k technology, *NASA Electronic Parts and Packaging Program, 2003.*

[8] A. McKerrow and P. Ho. Proceedings of the Low Dielectric Constant Materials and Interconnects Workshop, *SEMATECH* 1996, p. 199.

[9] Wen-Chang Chen, Shu-Chun Li, Bau Tong-Dai, Ming-Shih Tsai, Chemical Mechanical Polishing of Low-Dielectric-Constant Polymers: Hydrogen Silsesquioxane and Methyl Silsesquioxane, *Journal of The Electrochemical Society* 146 (8), 1999, p. 3004-3008.

[10] *International Technology Roadmap for Semiconductors.* Semiconductor Industry Association, 1994.

[11] C. Wyon, *Future technology for advanced MOS devices, Nuclear Instruments and Methods in Physics Research B* 186, 2002, p. 380-391.

[12] J. C. M. Garnett, *Philos. Trans. R. Soc.* London, Ser. B 203, 1904, p. 385.

[13] Paul S. Ho, Metrology Challenges for Ultra-Low-k Dielectrics for the 45 nm Node, *Future Fab Intl.* Volume 17 (online editions), 2004.

[14] W. Oh, Y. Hwang, Y.H. Park, M. Ree, S.-H. Chu, K. Char, J.K. Lee, S.Y. Kim, Optical, dielectric and thermal properties of nanoscaled films of polyalkylsilsesquioxane composites with star-shaped poly(ε -caprolactone) and their derived nanoporous analogues, *Polymer* 44, 2003, 2519-2527.

[15] Jun Xu a, J. Moxom, Shu Yang, R. Suzuki, T. Ohdaira, Dependence of porosity in methyl-silsesquioxane thin films on molecular weight of sacrificial triblock copolymer, *Chemical Physics Letters* 364, 2002, p. 309–313.

[16] Hae-Jeong Lee, Eric K. Lin, Howard Wang, Wen-li Wu, Wei Chen, Eric S. Moyer, Structural Comparison of Hydrogen Silsesquioxane Based Porous Low-k Thin Films Prepared with Varying Process Conditions, *Chem. Mater.* 14, 2002, p. 1845-1852 .

[17] Duan Li Ou, Pierre M. Chevalier, Iain A. Mackinnon, Katsuya Eguchi, Ronald Boisvert, Kai Su, Preparation of Microporous ORMOSILs by Thermal Degradation of Organically Modified Siloxane Resin, *Journal of Sol-Gel Science and Technology* 26, 2003, p. 407-417.

[18] Education pages, Cornell university, http://www.nanoindentation.cornell.edu, Alex A.Volinsky, Joseph B.Vella, William W.Gerberich, Fracture toughness, adhesion and mechanical properties of low-k dielectric thin films measured by nanoindentation, *Thin Solid Films* 429, 2003, p. 201–210.

[19] C.M. Flannery, C. Murray, I. Streiter, S.E. Schulz, Characterization of thin-film aerogel porosity and stiffness with laser-generated surface acoustic waves, *Thin Solid Films* 388, 2001, p. 1-4.

In: New Trends in Biochemical Physics Research
Editors: S. D. Varfolomeev et al., pp. 65-71
ISBN 1-60021-463-0

Chapter 7

OPTIMISED ANAEROBIC TREATMENT OF DIFFERENT KINDS OF ORGANIC WASTE IN A TECHNICAL SCALE DIGESTER

V. Vasilieva*[*]*, Ch. Resch*[†]*, R. Braun and R. Kirchmayr
Dep. IFA-Tulln, Institute for Environmental Biotechnology,
University of Natural Rasources and Applied Life Sciences, Vienna, Austria

ABSTRACT

Diversion of organic waste from land fill or incineration, to biological treatment by anaerobic digestion is of current interest in Austria. Now energy production from sewage sludge, solid wastes and energy crops by anaerobic digestion is wide spread practiced. The objective of this study was to optimise anaerobic treatment of different kinds of organic waste in a technical scale digester. New methods and laboratory analysis, for the description of microbiological process, were investigated and used. Detailed data have been collected from approximately 150 plants, which generate green electricity by means of energy crops, manure and organic waste. Based on the collected data from the biogas plants, a benchmark system for required process parameters was developed. This system helps the plant owners to operate their plants under monitored conditions in an economical way.

INTRODUCTION

The disposal of municipal solid wastes on wild or constructed landfills has been state of the art over decades. These waste management concept creates serious environmental and economic problems all over the world. These problems, together with price increases and

[*] viktoriya.vasilieva@boku.ac.at
[†] corresponding author: christoph.resch@boku.ac.at, Tel.:+43-2272-66280-514.

supply disturbances affecting petroleum products and natural gas, have led to research into alternative fuel and energy sources. Furthermore, in the developed countries the trend is to divert a large part of refuse away from landfill (Sans, 1995). According to the European landfill directive EC-99/31/EC waste must be treated before being landfilled to cut methane emissions. This regulation increases the pressure on the handling and treatment of organic wastes. Among biological treatments, anaerobic digestion is frequently the most cost-effective due to the high energy recovery linked to the process and it is limited environmental impact (Mata-Alvarez, 2000). The recycled residues can be used as fertilizer and soil conditioner. However, some wastes may lead to potential hygienic or contamination risk in agricultural land application (Mawdsley et al., 1995; Colleran, 1999; Sahlstreom, 2003; Westerman and Bicudo, 2005). Concurrently the European Union has set specific regulations in place to guarantee quality end products and to avoid risks for human or the environment. The most important regulations concerning the use of organic fertilizers are: EEC-Nitrates Directive (91/676/EEC), EEC-Sewage Sludge Directive: (86/278/EEC), EEC-Directive (2092/91) on organic production of agricultural products, ABP-Regulation EC (№ 1774/2002).

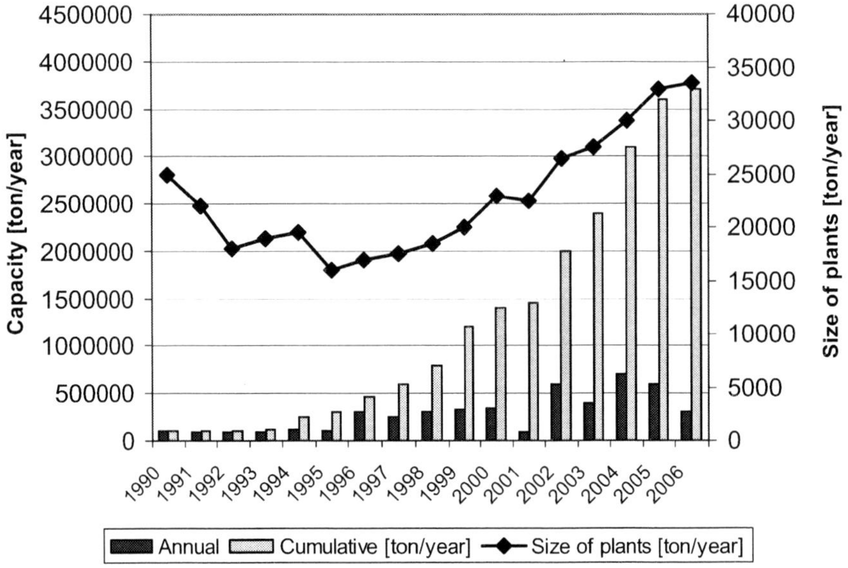

Figure 1. Installed Capacity in EU-15 for anerobic treatment of the organic fraction of municipal solid wastes (De Baere, 2005)

SITUATION IN EUROPE (EU-15)

The implementation of anaerobic digestion for the treatment of the organic fraction municipal solid waste (OFMSW) has been the major development in the field of waste treatment facilities in Europe during the last decade. As shown in Figure 1 more than a hundred and twenty waste treatment facilities have been constructed in Europe since 1990, in which a significant part of the organic waste is treated by anaerobic digestion. The capacity has greatly increased since 1990. Current evaluation of capacity indicates 124 plants with a total installed capacity of 3.905.000 ton per year by the end of 2006 in Eu-15 (Figure 1). This represents more than a doubling of the number of plants and almost a quadrupling of the capacity installed over a period of 6 years (De Baere, 2005). The growth in additional digestion capacity is increasing as well. This increasing growth is due to the number of plants combined with an increase in the size of the facilities.

PROCESS OPTIMIZATION

A typical anaerobic digestion plant for the treatment of OFMSW includes the following unit processes:

- Pretreatment: depending on the quality of the organic fraction several physical and mechanical pretreatment steps may be necessary. These are: shreddering, sieving, magnetic seperator and sometimes pulping.
- Input system for solid or liquid resources into the fermenter
- Fermenter: most systems for orgnic waste treatment are designed as completely stirred tank reactors (CSTR). The organic loading rate and required hydraulic retention time for the used substrates and the planned process temperature are the main design criteria. During this time the organic substances are metabolized by the microorganisms.
- The end products of an anerobic digestion process are valuable. The produced biogas has a methane content between 50 and 75 % and can be used in combined heat and power generatores to produce energy and heat or – after treatment – supplied into the natural gas grid. The liquid and solid digestion residues contain of certain amounts of nitrogen and phosporus are useful fertilizer. Large plants have to implement post treatment installation to improve the quality of the end-product.

It is very important to control all parameters in the full-technical scale fermenter to prevent malfunctions of the micobiological process. The chemical parameters such as pH, NH_4-N, TKN, COD, TS, TVS and volatile fatty acids (VFA)have to be monitored continuously.

Volatile fatty acids are important metabolites within the microbiological degradation process. Thus the VFA are widely used proposed as a process indicator (Björnsson, 2000). In a balanced anaerobic degradation process VFA do not accumulate over longer time period. Such an accumulation indicates an imbalance that will cause severe process inhibtion (Bryant, 1979). In particular high concentrations of propionic and acetic acids lead to very low methane productivity and process instability. Since VFA can be measured exactly in the laboratory they are the best known process control parameters these days. They allow to have a better view into the degradation process but VFA cannot assess the activity or kinetics of the microbial community.

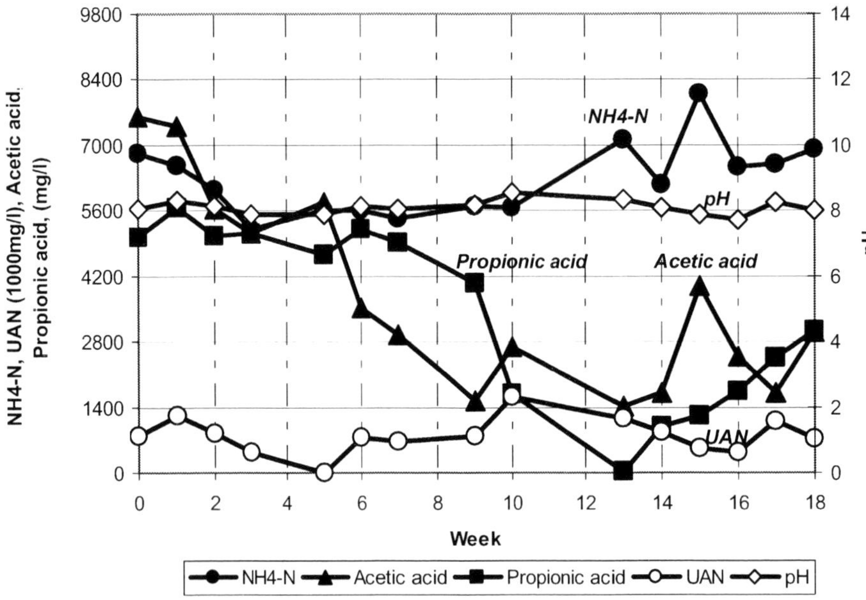

Figure Figure 2 shows how VFA can be used to monitor a full scale anaerobic digestion system. There the analytical monitoring of the fermentation process of a biogas plant in Austria over an 18 week period is shown. This biogas plant had a very low methane productivity.

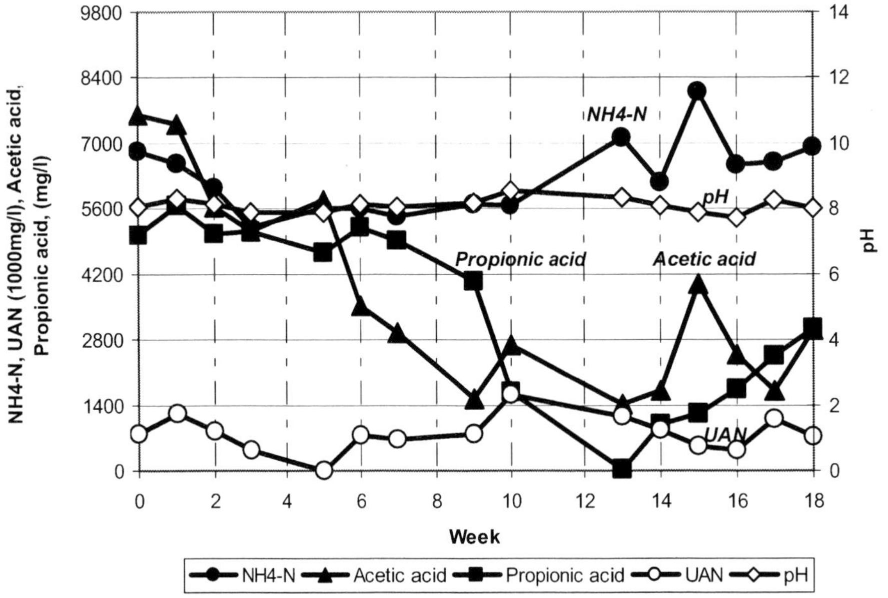

Figure 2. Analytical monitoring of an Austrian plant over 18 weeks period. Restart of an imbalanaced process

The VFA were measured in very high concentrations and the methane production was inhibited severely. To stabilize the process it was decided to decrease the organic waste input. As expected the concentration of acetic acid decreased rapidly to a reasonable level within 4 weeks. During that time even a slight increase of propionic acid concentration was observed. It took longer, approximately 8 weeks, until the concentration of the propionic acid decreased. The odd-numbered (C3) propionic acid is harder to degrade because its decomposition is only possible in a small termodynamical range. As the propionic acid cannot be split into two acetic acid molecules, this results in the longest digestion time for all VFAs (Schink, 1997, Cord-Ruwisch et.al., 1997). But not only VFA concentrations influence the microbiological process stability: changes of the uniionized dissociation equvilant of soluble ammonium (UAN) concentration were expected due to slight increase of pH as a reaction of VFA degradation. The UAN concentration increases with an increase in pH and temperature. UAN inhibts the microorganisms directly after exceeding a certain concentration level, different values are reported (Koster, 1986, Koster, 1988, Angelidaki, 1994). This example illustrates that a significant number of process parameters interact together and have be to monitored coinstantanuously.

Within 10 weeks of substrate input reduction the process was stabilized and the input of the organic sources proceeded under normal conditions. More than two months of monitoring and optimization were required to make the biogas plant operating properly again. Such imbalaces in the process for an extended time can be expensive and unfavourable for plant owners.

Benchmark System

Based on the collected data from 150 Austrian biogas plants, a benchmark system for required process parameters was developed. This system does not quote strict concentration limits for several parameters. It is divided into three category groups for each parameter (good, bad and alarm condition) and offers operation ranges for the plant operators. Thus faults in the process may be detected early and corrected in a reliable way.

Draw up Plans

In future, other methods to describe the microbiological process will be developed and adapted, such as:

- bioluminescence method, based on the measurement of Adenosine Triphosphate (ATP).The energy obtained is stored in ATP, which is then transferred within the organism to perform work functions, thus enabling the measurement of the stability and productivity of the anaerobic digestion process ;
- light fluorescence microscopy method, achieves a level of sophistication that far exceeds simple observation by the human or electronic eye and allows appraise of quality and quantity of *archae bacteria*;

- near infra red spectroscopy method, with a low number of wavelengths show the general process status and could be used as a predictor for control strategies.

Conclusions

Anaerobic digestion is a forward-looking technology for the treatment of most high concentrated organic wastes. Almost all kinds of organic wastes are suitable for anaerobic treatment. Both outputs of this process are beneficial. The methane gas is a primary energy source that can be supllied in the natural gas grid or can be used in combined heat and power generators to produce green electricity and heat. The solid and liquid process residues can be used as valuable fertilizers. The methane yield mainly depends on the substrate used and on the careful operation of the process. The microbiological process in the fermenters should be controlled constantly. In future new process monitoring tools will be available. These methods should be applicable for the plant operatores on-site and should allow a deeper view into the activity and kinetics of the microorganisms.

References

Angelidaki I., Ahring B. K., Anaerobic thermophilic digestion of manure at different ammonia loads: effect of temperature, *Water Res.,* 1994, 28(3), 727-731.

Björnsson L. (2000). Intensification of the biogas process by improved process monitoring. PhD thesis, Department of Biotechnology, Lund University, Sweden.

Bryant M. P. (1979). Microbial methane production-theoretical aspects. *Journal of Animal Science,* 193-201.

Colleran E. (1999). Hygienic and sanitation requirements in biogas plants treating animal manures or mixtures of manures and other organic wastes. In: *The European Anaerobic Digestion Network, www.adnett.org.25.1.2000.*

Cord-Ruwisch R., Mercz T.I. and Strong G.E. (1997). Dissolved Hydrogen Concentration as an On-Line Control Parameter for the Automated Operation and Optimization of Anaerobic Digesters. *Biotech. Bioeng.* 56(6), 626-634.

De Baere L. (2005). Will anaerobic digestion of solid waste survive in the future? In: *Proceedings of the 4th International symposium "Anaerobic digestion of solid waste"* in Copenhagen, August 31-September 2, 2005,72-81.

EEC-Directive, 2092/91 on organic production of agricultural products and indications referring thereto on agricultural products and foodstuffs to include livestock production

EEC-Nitrates Directive: COUNCIL DIRECTIVE of 12 December 1991 concerning the protection of waters against pollution caused by nitrates from agricultural sources (91/676/EEC)

EEC-Regulation, 1774/2002 of the European Parliament and the council from the 3rd October 2002 laying down health rules for animal by-products not intented for human consumption. *Official Journal of the European Communities* L 273, I-95. 10-10-2002

EEC-Sewage Sludge Directive: Council Directive 86/278/EEC of 12 June 1986 on the protection of the environment, and in particular of the soil, when sewage sludge is used in agriculture.

http://euopa.eu.int/erur-lex/pri/de/ojdat/2002/l_273/l_27320021010de00010095.pdf

Koster I. W., Koomen E., Ammonium inhibition of the maximum growth rate of hydrogenotrophic methanogens at varios pH-levels and temperatures, *Microbiol. Biotechnol., 1988*, 28, 500-505.

Koster I: W., Characteristics of the pH-influenced adaption of methanogenic sludge to ammonium toxicity, *J. Chem. Tech. Biotechnol.,* 1986, 36, 445-495.

Mata-Alvarez J., Mace S. and Llabres P. (2000). Anaerobic digestion of organic wastes. An overview of research achievements and perspectives. *Biores. Technol.* 74,3-16.

Mawdsley J., Bardgett R., Merrt R., Pain B. and Theodorou M. (1995). Pathogens in livestock waste, their potential for movement through soil and environmental pollution. *Appl. Soil Ecol.,*2, 1-15.

Sahlstroem L. (2003). A review of survival of pathogenic bacteria in organic waste used in biogas plants. Review paper. *Biores.Technol.*, 87, 161-166.

Sans C., Mata-Alvarez J., Checci F., Pavan P. and Bassetti A. (1995). Volatile fatty acids production by mesophilic fermentation of mechanically sorted urban organic wastes in a plug-flow reactor. *Biores.Technol.*, 51, 89-96.

Schink B. (1997). Energetics of Syntrophic Cooperation in Methanogenic Degradetion. *Microbiology and molecular biology reviews*, Vol. 61, no. 2, 262-280.

Westerman P. and Bicudo J. (2005). Management considerations for organic waste use in agriculture. *Biores. Technol.*, 96, 215-221.

In: New Trends in Biochemical Physics Research
Editors: S. D. Varfolomeev et al., pp. 73-82
ISBN 1-60021-463-0

Chapter 8

Regulatory Proteins Biologically Active in Ultralow Doses from Mammalian Glands and Their Secretions

***P. A. Nazarova*[*], *V. P. Yamskova*[*], *M. S. Krasnov*[*], *A. G. Filatova*[***] and *I. A. Yamskov*[**]**

[*]Kol'tsov Institute of Developmental Biology
26 Vavilova str., Moscow 117334, Russia, Russian Academy of Sciencis
[**]N.A. Nesmeyanov Institute of Organoelement Compounds,
28 Vavilova str., Moscow 117813, Russia, Russian Academy of Sciencis
[***]N.M. Emanuel Institute of Biochemical Physics,
Russian Academy of Sciencis

Abstract

Proteins exhibiting biological activity in ultralow doses, or regulatory proteins (RPs) were isolated from the bovine prostate, mammary gland, and milk. Using an immunohistochemical method, these RPs were localized in the intercellular spaces of the secretory epitheliums of the prostate and lactating mammary gland. Experiments with the model of *in vitro* roller culture of the murine prostate showed that prostate-derived RPs in ultralow doses had an effect on the production of secretion by glandular cells.

Keywords: Regulatory proteins, ultralow doses, prostate, milk, mammary gland, secretion, roller organ culture

Introduction

Regulatory proteins (RPs) exhibiting biological activity in ultralow doses (10^{-8}–10^{-12} mg protein/ml) have been detected in many tissues of vertebrates. Experimental data confirm that proteins of this group have a regulatory effect on cell adhesion, migration, proliferation, and differentiation, as well as on the functioning of major enzyme systems and protein synthesis [7–11]. In this study, we made an attempt to identify RPs in mammalian glands and their secretions and to analyze their physicochemical properties and biological activity. The bovine prostate, mammary gland, and whole milk were chosen as prospective sources of RPs. The mammary gland and milk offered an opportunity to study RPs of the gland itself separately from RPs of its secretion. In the case of prostate, the gland and its secretion were studied together.

Such a study appears relevant in view of the exclusively important role of milk and diary products in the human diet, especially at an early age. In addition to biologically active substances identified to date, milk is likely to contain some unknown components that also account for the normal development of individuals in the postnatal period. The milk-derived RP biologically active at ultralow doses (see below) may belong to this category.

No less important is the role of the prostate in the reproductive function. Its secretion, which is a specific medium providing for high motility and fertility of spermatozoa [2], has a unique composition. In particular, it contains Zn-α-2-glycoprotein (34 kDa) [3], proteolytic enzymes, prostaglandins, the prostate-specific antigen (PSA, 29 kDa) [2], the prostate secretory protein PSP94 (13 kDa) [4], and specific vesicles named prostasomes [5, 6]. The functions of many components of prostatic secretion are as yet poorly known, and the same concerns biological agents regulating the functioning of the prostate. The results of this study show that RPs isolated from the bovine prostate have a specific stimulating effect on the synthesis of prostatic secretion components.

EXPERIMENTAL

Milk (5 l) was collected 1 h after the milking of healthy cows on the Kommunarka State Farm (Moscow Region), in summer. The prostate glands of mature bulls (60 specimens) and mammary glands of barren cows (2 specimens) were from meat-packing plants in Ramenki and Sergiev Posad (Moscow Region), respectively. All reagents were of chemical purity grade. Solutions were made in deionized water (15 MΩ) from a MilliQ column.

The extracts from the prostate and mammary gland were prepared by the early developed method [7]. Milk and tissue extracts were treated with ammonium sulfate and centrifuged to remove ballast protein [8]. The resulting preparations were separated by isoelectrofocusing (IEF) [8]. In a special experiment, whole milk (10 ml) was separated. IEF was performed in an LKB-110 column (LKB, Sweden) in a sucrose density gradient with ampholytes (pH 2.5–10, LKB) at 500–1500 W for 96 h, at 4°C [16].

Protein concentration was measured by the colorimetric method with bicinchoninic acid, at 562 nm [17].

To determine the molecular weight of RPs, the fractions obtained after IEF were resolved by electrophoresis in 15% PAAG in the presence of 0.1% sodium dodecylsulfate, as described [12]. Electrophoresis was performed in a Mini-Protean III cell (BioRad, USA). A molecular weight marker kit (LKB, Sweden) used as a reference included the following proteins: phosphorylase b (94 kDa), bovine serum albumin (67 kDa), ovalbumin (45 kDa), carbonic anhydrase (30 kDa), soybean trypsin inhibitor (20 kDa), and chicken egg white lysozyme (14.4 kDa).

To obtain RPs in a biologically active form, electrophoresis under nondenaturing conditions was performed [12]. After electrophoresis, the zone corresponding to proteins with $R_f = 0.8–1.0$ was cut out, and proteins were eluted by incubating it in several small volumes of water. The eluates were pooled, dialyzed against water, and lyophilized.

Fractions obtained at each stage of purification were checked for biological activity using a cell adhesion test developed for biological indication of RPs [9]. Experimental data were processed by methods of variation statistics. Student's *t*-test was used to estimate significance of differences.

Antisera for immunohistochemical identification of RPs were obtained by immunizing Chinchilla rabbits weighing 1.5–2 kg with weakly acidic RP fractions (pI 4.5–5.6). The antigen solution (0.2 mg/ml) was mixed with the equal volume of complete Freund's adjuvant, and each rabbit received 1 ml of the mixture injected under the skin on the back and into the femoral muscles at approximately equal doses. Thereafter, injections of the antigen mixed with incomplete Freund's adjuvant (1 : 1) were made in the same way at seven-day intervals. Blood samples taken from the marginal ear vein beginning from the fourth week after the first injection were tested for the presence of specific antibodies by means of ELISA [10, 13].

Experiments on immunohistochemical localization of RPs were performed with histological sections of the lactating rat mammary gland and the murine prostate. The sections were treated with primary antibodies obtained as described above and secondary FITC-labeled antibodies to rabbit immunoglobulins [14].

To estimate parameters of the secondary structure of test proteins, their circular dichroism spectra were recorded in a Jasco 720 dichrograph (Japan) equipped with a computer data processing system, as described [15]. The spectra were analyzed using the k2d program [17].

An analysis of RP preparations by the method of atomic power microscopy was performed using a FEMTOSCAN ONLINE instrument with the FEMTOSCAN program for data processing (Center of Advanced Technologies, Moscow).

RP preparations from the bovine prostate were tested for biological activity on roller cultures of the murine prostate. In these experiments, F1 hybrid mice were used. The upper lobe of the prostate was surgically removed and placed in a vial with the culture medium (90% medium 199 and 10% embryonic fetal calf serum with 80 mg/ml gentamicin sulfate). The test cultures were supplemented with the acidic (pI 1.7–3.0) and weakly acidic (pI 4.5–5.6) RP preparations from the bovine prostate to the final concentrations of 10^{-10} and 10^{-9} mg protein/ml. The vials were installed in an RM 5 roller (Assistant, Germany) and incubated at 35 rpm and 37°C for 24 h. The cultures were fixed with the Bouin solution. Histological sections (5 μm) were stained with hematoxylin–eosin and examined under a light microscope.

Results and Discussion

To identify RPs exhibiting biological activity at ultralow doses in milk and extracts from the bovine prostate and mammary gland, we used the procedure tested in our previous studies [8], which involved two rounds of protein precipitation with ammonium sulfate and IEF separation of proteins that remained soluble in a saturated salt solution. With regard to our previous data, the fractions of acidic (pI 1.5–3.0) and weakly acidic (pI 4.5–5.6) proteins were collected and checked for the presence of RPs in biological tests [8, 16].

As follows from Table 1, both these fractions from the prostate extracts proved to be biologically active in ultralow doses. In the case of milk, only the weakly acidic fraction was active, whereas the acidic fraction lacked biological activity (the same was true for IEF fractions of whole milk). In the extract from the mammary gland, conversely, only the acidic fraction was active.

These results suggest that the weakly acidic RP detected in milk and the prostate may be identical to the RP previously identified in blood serum, whose isoelectric point is in almost the same range (pI 4.5–5.1) [8, 15]. Published data show that some proteins can pass from the blood serum into secretions of the glands studied [1, 3], which may be true of the above RP. This aspect is interesting in view of the data that the blood serum RP, being active during early embryonic development, loses its activity by the moment of birth and remains inactive (bound to an inactivator protein) throughout postnatal life, so that its function in adults is unknown [8, 15]. It is logical to assume that the blood serum RP can pass into milk when it dissociates from the inactivator protein and recovers its biological activity, the more so that milk has unique nutritive properties and is necessary for the normal development and intensive growth of the progeny during the first year of life. Our data that whole milk probably contains this RP in the active state provide evidence in favor of this assumption.

Table 1. Parameters of RP preparations tested for biological activity

Source	Fraction collected after IEF, pI range	Biologically active concentration, mg/ml*
Milk after ammonium sulfate treatment	4.5–5.6	10^{-10}
Whole milk	4.5–5.6	10^{-9}
Mammary gland extract after ammonium sulfate treatment	1.5–3.0	10^{-10}
Prostate extract after ammonium sulfate treatment	1.5–3.0	10^{-10}
	4.5–5.6	10^{-14}

*In the cell adhesion test.

Taking into account that the blood serum RP stimulates cell adhesion and proliferation [8], the question arises as to whether it passes from blood serum into milk via a transcellular or paracellular route [18] or is synthesized by the mammary gland during lactation. This question will be elucidated in our subsequent studies.

The weakly acidic RP preparations from the prostate extract and milk (pI 4.5–5.6) were resolved by SDS-electrophoresis in 15% PAAG. The results showed that the main component of each preparation was represented by a low-molecular-weight protein fraction (< 10 kDa) (Figures 1, 2).

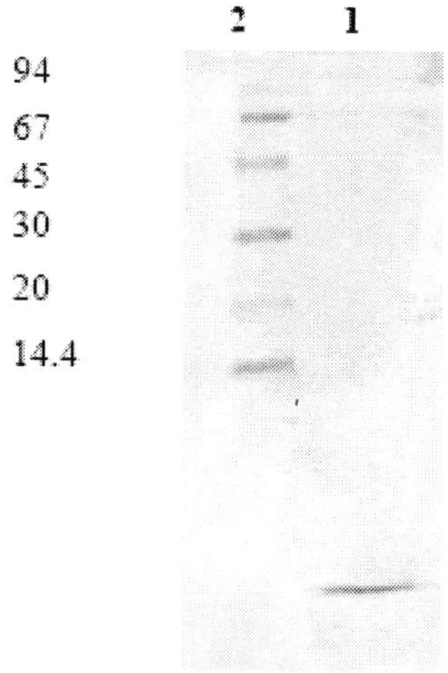

Figure 1. Electrophoretic pattern of weakly acidic protein fraction (pI 4.5–5.6) from milk in 15% PAAG: (*1*) milk protein; (*2*) molecular weight markers, kDa.

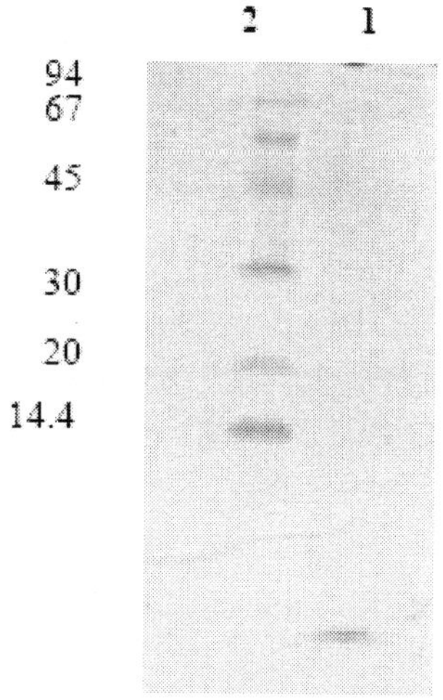

Figure 2. Electrophoretic pattern of weakly acidic protein fraction (pI 4.5–5.6) from the prostate in 15% PAAG: (*1*) prostate protein; (*2*) molecular weight markers, kDa.

To obtain the corresponding proteins in a biologically active form, the same RP preparations were separated by electrophoresis under denaturing conditions, and low-molecular-weight fractions were eluted from the gel. They were used as antigens for immunizing rabbits. By means of immunohistochemical analysis with antisera to RPs from the bovine prostate extract and milk, the corresponding low-molecular-weight proteins were localized in the intercellular spaces of glandular epitheliums of the murine prostate and the lactating rat mammary gland, respectively (Figures 3, 4).

The preparations obtained after nondenaturing electrophoresis were also used to analyze the secondary structure of RPs by the method of circular dichroism. The results showed that the polypeptide chains of these RPs consisted mainly of β-sheet regions and structures described in terms of stochastic globule, which was in full agreement with data on the blood serum RP [15] (Table 2). It should be noted that the prevalence of β-sheet regions in the folding pattern implies that these proteins should have a tendency to form intermolecular aggregations [19]. This was confirmed by the results of atomic power microscopy, which showed that, irrespective of dilution, RPs from the blood serum and milk occurred in solution in the form of spherical particles with a diameter of 100 nm and greater (Figure 5).

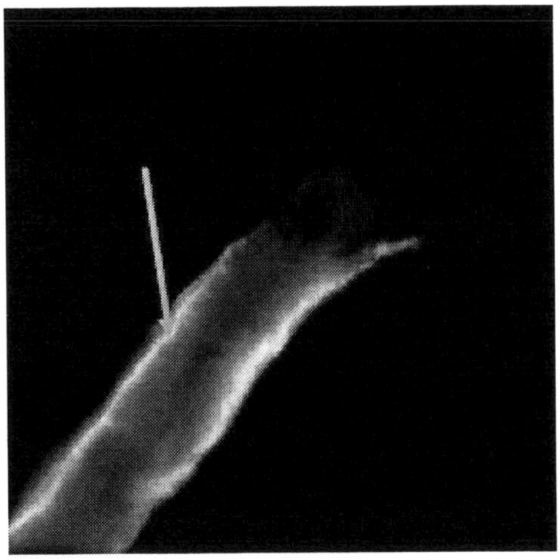

Figure 3. Immunohistochemical localization of the weakly acidic RP from milk. in a histological section of the lactating rat mammary gland. The protein is localized in the duct of rat mammary gland (arrow). Magnification: ocular 10×, objective lens 40×.

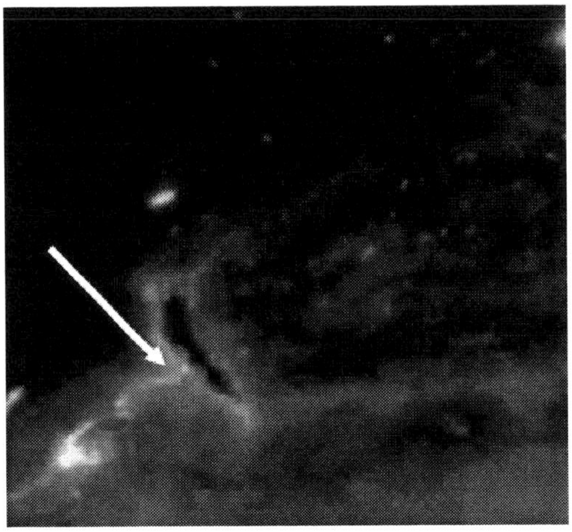

Figure 4. Immunohistochemical localization of the weakly acidic RP from the bovine prostate in a histological section of the murine prostate. The protein is localized in the duct of murine prostate (arrow). Magnification: ocular 10×, objective lens 40×.

Table 2. Parameters of the secondary structure of weakly acidic RPs from milk, prostate, and blood serum

Elements of secondary structure	Content in RP molecules, %		
	milk RP	Prostate RP	Blood serum RP
α-Helix	6.2 ± 0.5%	6.4 ± 0.5%	10.0 ± 0.5%
Antiparallel β-sheet	47.7 ± 0.5%	47.1 ± 0.5%	38.0 ± 0.5%
Parallel β-sheet	3.7 ± 0.5%	3.7 ± 0.5%	2.0 ± 0.5%
β-Fold	16.8 ± 0.5%	14.1 ± 0.5%	18.0 ± 0.5%
Random coil	25.6 ± 0.5%	28.7 ± 0.5%	32.0 ± 0.5%
Total	100 ± 0.5%	100 ± 0.5%	100 ±0.5%

The biological activity of RP preparations from the bovine prostate was tested on the model of roller organ cultures of the murine prostate. The effects of acidic and weakly acidic RP fractions in ultralow doses were tested separately. In cultures treated with the acidic fraction, intensified production of homogeneous secretion by glandular cells was observed. This was accompanied by mass cell mortality (more than 90%), disturbances of cell adhesion, karyopycnosis, and shrinking of epithelial cells, although the lobular structure of the gland was partially preserved (Figures 7a–7c). In control cultures, the tissue looked more viable, especially in the peripheral zone. The structure of individual ducts with epithelial trabeculae was largely preserved, but no secretory activity was observed (Figures 6a–6c).

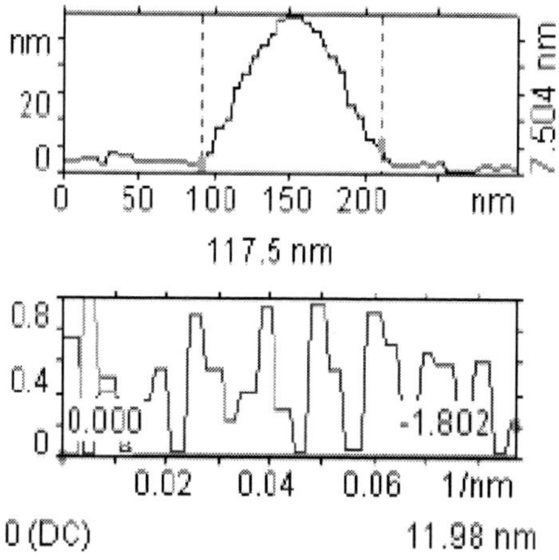

Figure 5. Size of nanoparticles formed by weakly acidic milk RP in solutions.

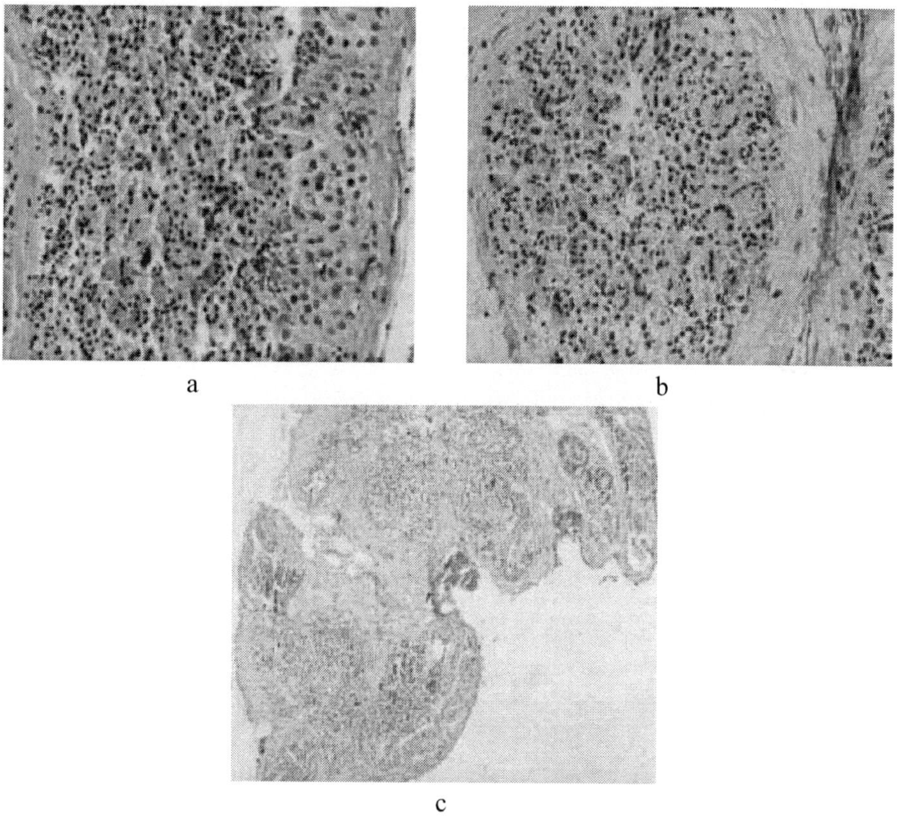

Figure 6. Transverse section of the murine prostate cultured for 24 h without adding RP preparations (control). Magnification: (a, b) ocular 10×, objective lens 20×; (c) ocular 10×, objective lens 4×.

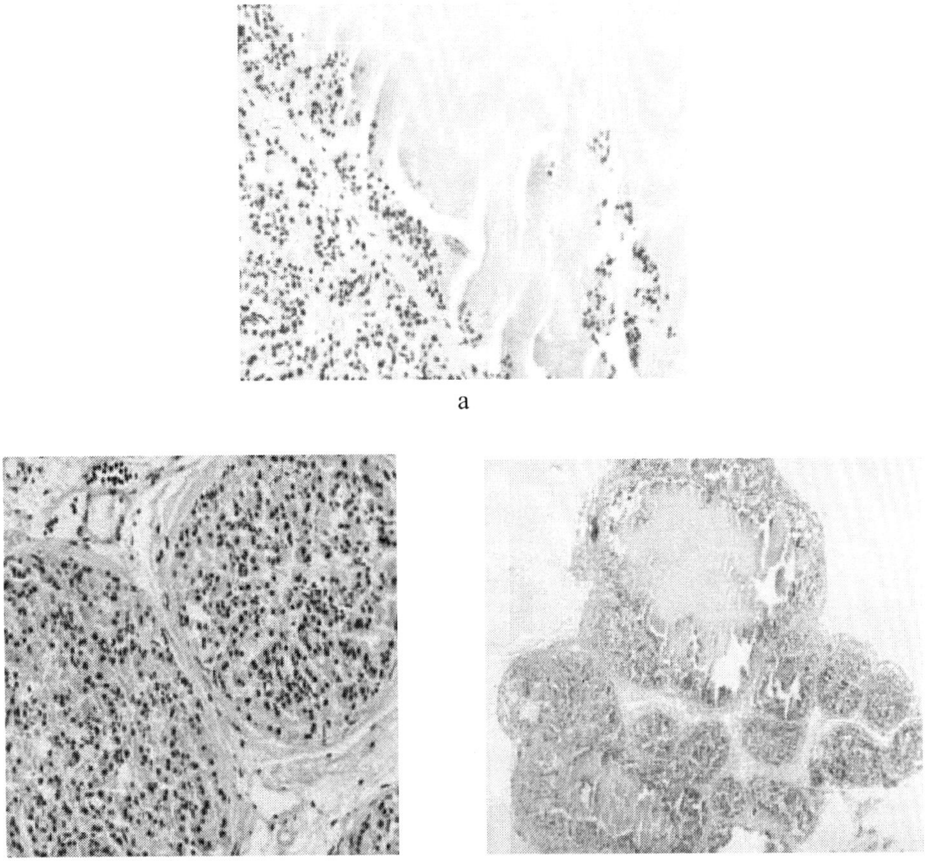

a

b

c

Figure 7. Transverse section of the murine prostate cultured for 24 h with the acidic RP fraction from the bovine prostate (10^{-10} mg/ml). Magnification: (a, b) ocular 10×, objective lens 20×; (c) ocular 10×, objective lens 4×.

In cultures treated with the weakly acidic RP preparation, cell mortality was considerable. Connective tissue septa between individual ducts were damaged, and the gland acquired a homogeneous structure. Despite cell death, the trabecular structure of the ducts was partially preserved, and adhesive cell-to-cell interactions were well manifested. However, many epithelial trabeculae died off because of excessive secretory activity, and all spaces between their remains were filled with the secretion (Figures 8a–8c). It is noteworthy that the secretion produced in response to the acidic and weakly acidic RPs differed in texture: it was homogeneous in the former case (Figure 7a) and granular in the latter case (Figure 8a).

These results show that ultralow doses of the acidic and weakly acidic RPs from the bovine prostate stimulate secretory cells of the murine prostate, which is manifested in the synthesis of secretion in much greater amounts that in the control. A probable explanation to changes observed at the histological level is that the stimulating effect of these RPs is so strong that the intensive secretory activity exhausts epithelial cells and many of them die after releasing the secretion into the gland duct.

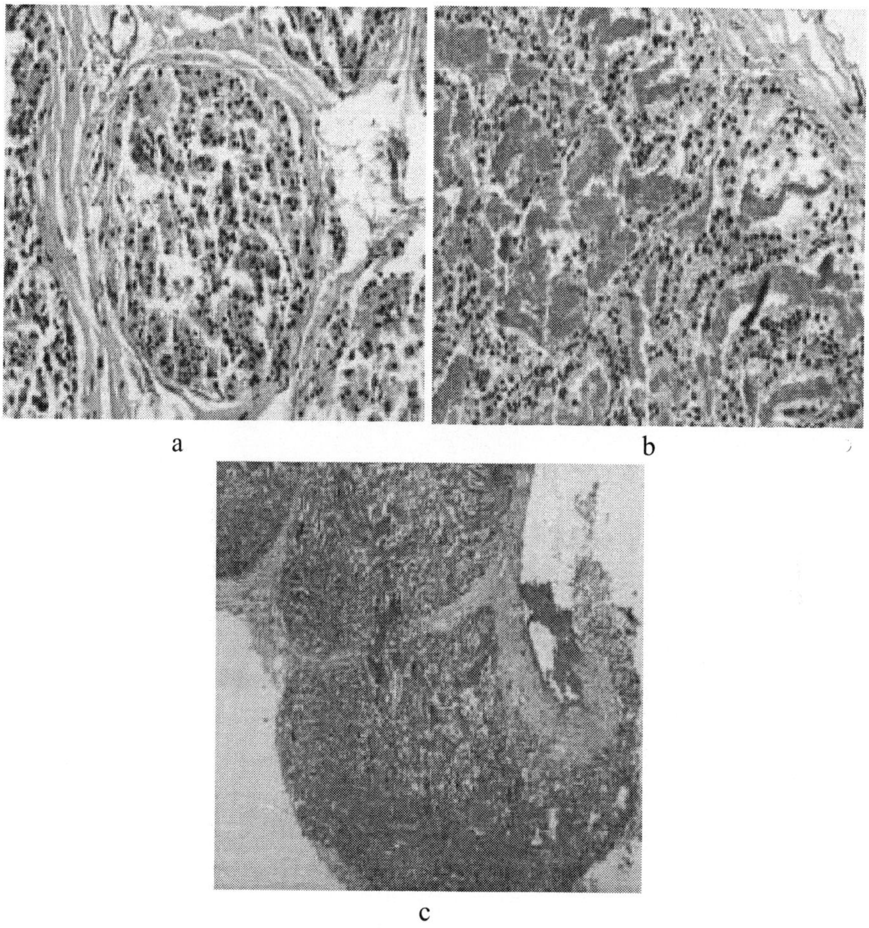

Figure 8. Transverse section of the murine prostate cultured for 24 h with the weakly acidic RP fraction from the bovine prostate (10^{-14} mg/ml). Magnification: (a, b) ocular 10×, objective lens 20×; (c) ocular 10×, objective lens 4×.

Thus, the bovine prostate, mammary gland, and milk contain low-molecular-weight RPs that exhibit biological activity in ultralow doses. Their secondary structure is characterized by the prevalence of β-sheet and stochastic globule regions. In solutions, these RPs are found in the form of spherical nanoparticles. It has been proposed that the weakly acidic RP from the prostate extract and milk is identical to the RP previously identified in the blood serum. The RP fractions from the bovine prostate stimulate the secretory activity of glandular cells in cultures of the murine prostate.

REFERENCES

[1] K.K. Gorbatova, *"Biochemistry of Milk and Dairy Products"*, St. Petersburg, GIORD, 2004, 320 pp. (in Russian)

[2] O.L. Tiktinskii, "A Manual of Andrology", Leningrad, *Meditsina*, 1990, 312 pp. (in Russian).

[3] Y. Kim, L. Fung, G. Michael, G. Spencer, M. Duncan, *The Prostate,* Vol. 61, pp. 171–181, 2004.

[4] N. Shukeir, A. Arakelian, S. Kadhim, S. Garde, S. Rabbani, *Cancer Research*, Vol. 63, N 9, pp. 2072–2078, 2003.

[5] G. Angelita, C. Eugene, X. Tao, P. Shannon, J. White, D. Goodlett, L. Hood, B. Lin, *The Prostate,* Vol. 56, pp. 150–161, 2003.

[6] G. Arienti, E. Carlini, C. Saccardi, C. Palmerini, *Journal of Cellular and Molecular Medicine*, Vol. 8, N 1, pp. 77–84, 2004.

[7] V.P. Yamskova, E.A. Modyanova, M.M. Reznikova, A.G. Malenkov, *Molekulyarnaya Biologiya,* Vol. 11, N 5, pp. 1147–1154, 1977.

[8] V.P. Yamskova, M.M. Reznikova, *Zhurnal Obshchei Biologii,* Vol. 52, N 2, pp. 181–191, 1991.

[9] V.P. Yamskova, N.V. Nechaeva, N.B. Tumanova, Yu.G. Yurovitskii, T.E. Novikova, V.I. Fateeva, I.G. Gvazava, *Izvestiya Rossiiskoi Akademii Nauk. Seriya Biologicheskaya*, N 2, pp. 190–196, 1994.

[10] M.S. Krasnov, E.N. Grigoryan, V.P. Yamskova, Izvestiya Rossiiskoi Akademii Nauk. *Seriya Biologicheskaya*, N 1, pp. 22–36, 2003.

[11] M.S. Krasnov, D.V. Margasyuk, I.A. Yamskov, V.P. Yamskova, *Radiatsionnaya Biologiya. Radioekologiya,* Vol. 43, N 3, pp. 269–272, 2003.

[12] U.K. Laemmli, *Nature*, Vol. 227, N 5259, pp. 680–685, 1970.

[13] *"Immunoassays: A Practical Approach",* Ed. by J.P. Gosling, Oxford University Press, 2000, 215 pp.

[14] N.V. Engelhardt, "Immunochemical Analysis", Moscow, *Meditsina*, 1968, 346 pp. (in Russian).

[15] I.A. Yamskov, A.A. Vinogradov, A.N. Danilenko, L.A. Maslova, E.Yu. Rybakova, V.P. Yamskova, *Prikladnaya Biokhimiya i Mikrobiologiya,* Vol. 37, N 1, pp. 36–42, 2001.

[16] I.A. Yamskov, V.P. Yamskova, A.N. Danilenko, Z.S. Klemenkova, B.G. Antipov, F.R. Chernikov, M.M. Gusynina, E.Yu. Rybakova, *Rossiiskii Khimicheskii Zhurnal,* Vol. 43, N 5, pp. 34–39, 1999.

[17] S.Yu. Venyaminov, J.T. Yang. "Determination of protein secondary structure", pp.69-107, in the book *"Circular Dichroism and the conformational analysis of Biomolecules"*, Ed. by G.D. Fasman, New York, Plenum Press., 1996.

In: New Trends in Biochemical Physics Research
Editors: S. D. Varfolomeev et al., pp. 83-87
ISBN 1-60021-463-0

Chapter 9

THERMAL DEGRADATION AND COMBUSTIBILITY OF NANO-FILLED POLYPROPYLENE COMPOSITION WITH MAGNESIUM HYDROXIDE

A. D. Rakhimkulov and S. M. Lomakin
Institute of Biochemical Physics of Russian Academy of Science,
4, Kosygin str., Moscow 119991, Russia

ABSTRACT

The flammability characteristics of PP, PP nanocomposite based on layered silicate (Cloisite 20A), composition of PP with MH (50% wt.) and the composition of PP nanocomposite with 20 % wt. of MH were examined with a cone calorimeter at an external heat flux of 35 kW/m^2. The peaks heat release rate of PP/ MH (20% wt.) and PP–MPP-Cloisite 20A (5.6% wt.) with MH (20% wt.) decrease by a factor of six and four, respectively, in comparison with pristine PP. It means that relatively small incorporations of MH to PP nanocomposites can substantially reduce the important factor of combustibility, such as maximum rate of heat release. The main purpose of the present work was the compare the flame retardancy of PP, PP/ MH (20% wt.) and PP–MPP-Cloisite 20A (5.6% wt.) with MH (20% wt.) in order to obtain much better flame resistance using a small loading of MH.

Keywords: Combustion, flame retardant, magnesium hydroxide, nanocomposite, polypropylene, rate of heat release, thermal degradation

Thermal degradation and combustion of polypropylene (PP), flame retardant (FR) compositions of polypropylene with magnesium hydroxide (MH) and nanocomposites of PP based on layered silicates are widely studied and discussed previously [1-5].

It is well-known that PP degrades completely without solid residue in inert atmosphere and in air. The mechanism of PP thermal degradation presents the radical chain process which occurs by random mechanism mostly. Figure 1 shows the typical chromatograms of PP pyrolysis products in inert atmosphere. Analysis of the main decomposition products indicates that the process of radical chain degradation leads to formation of alkanes, alkenes and dienes C3-C31 [6].

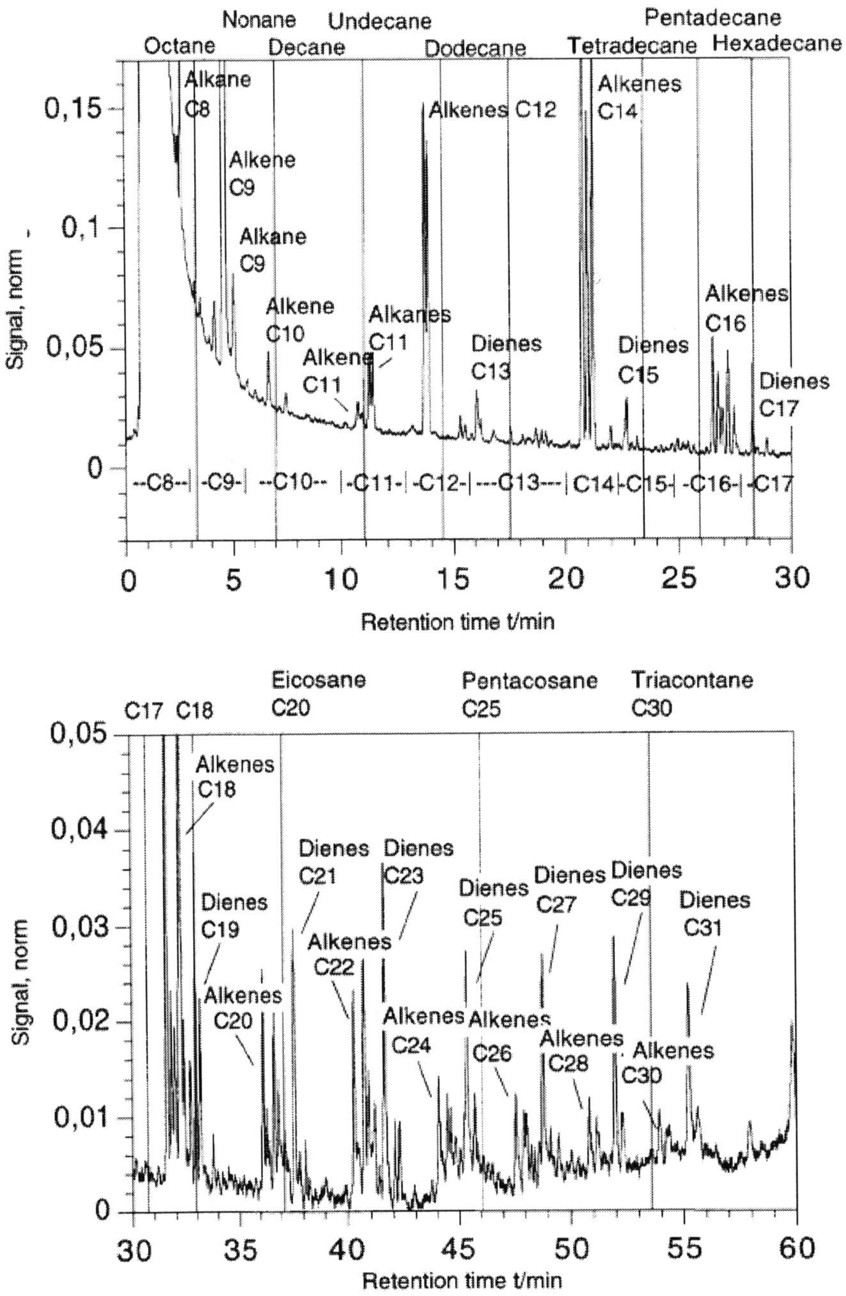

Figure 1. Chromatograms of PP pyrolysis products in inert atmosphere at 450°C [6].

In the presence of oxygen the process of PP thermal degradation is initiated by the early formed hydroperoxides' (HP) decomposition followed by scission of the main polymer chain as well as branched radical reactions. In this case the oxygen-containing products (alcohols, aldehydes and ketones) have been detected in the products of PP pyrolysis as well as alkanes, alkenes and dienes. At temperatures above 400°C in inert atmosphere and in air the overall process of PP degradation is described by random mechanism. That is why, a large amount of hydrocarbons is found at the temperatures above 400°C with/without of the presence of oxygen. Hydrocarbons present a major threat in the combustion of main plastics.

A large number of traditional FRs is used for PP in industry applications: halogenated FRs, phosphorous containing FRs, metal hydroxides (i.e. MH) [7]. Today MH is one of the commercially well-spread and ecologically friendly FR for PP. The mechanism of action of MH is based on dilution of gaseous flame zone by the water vapor and its endothermic effect in solid phase.

$$Mg(OH)_2 \xrightarrow{T>340°C} MgO + H_2O$$

Recently Shehata studied the combustibility of PP with MH in cone calorimeter [8]. Figure 2 presents the heat release rate vs. time under the incident heat flux of 35 kW/m^2 in cone calorimeter study for PP and compositions of PP with MH. It can be seen that the maximum heat release rate of PP is about three times larger than the one for the composition of PP with MH. It should be noted, that more than 50% MH by mass in composition with PP needed for desirable flame retardant effect. Obviously, this high loading of MH leads to decrease of mechanical properties of polymeric composition.

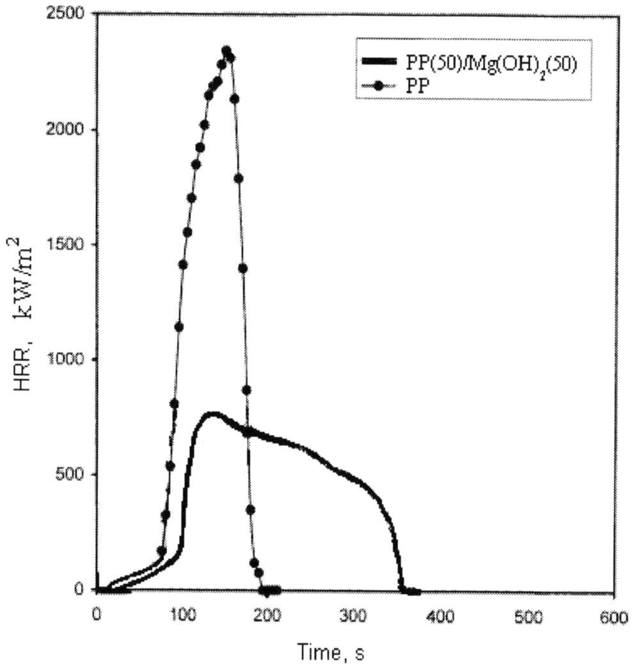

Figure 2. Heat release rate vs. time at the incident heat flux of 35 kW/m^2 in cone calorimeter study for PP and compositions of PP with MH [8].

The modern approach to the enhancement of thermal stability and fire resistance based on the use of polymer nanocomposites has been extensively developed in the last years [9–12]. Recently, the study of flame retardant properties of PP nanocomposites based on layered silicates showed the loadings of only 4-7% by mass of nanosilicates to the PP composition yield in more than 100% decrease of the maximum heat release rate [10-12].

Previously, there were discussed the basic features of the thermal degradation and combustion of PP nanocomposites based on organically modified alumina silicates [13]. The main reason for the reduction in the flammability of the PP nanocomposite is the formation of heat resistant char on the surface of the burning material.

The flammability characteristics of PP and its nanocomposite based on layered silicate (Cloisite 20A) were examined with a cone calorimeter at an external heat flux of 35 kW/m^2 for the samples having a standard surface area of 70×70 mm and identical masses of 14.0±0.1 g. Figure 3 depicts the plots of the basic ignitability characteristic - heat release rate, versus time for PP (1) , as well as for the PP-MPP (low molecular maleated PP) composition with 7 wt % Cloisite 20A (3) and PP with 7 wt % of Cloisite 20A nanocomposites (2). The last sample differs from the PP–MPP– Cloisite 20A nanocomposite in that it lacks maleated PP and represents a completely intercalated PP nanocomposite.

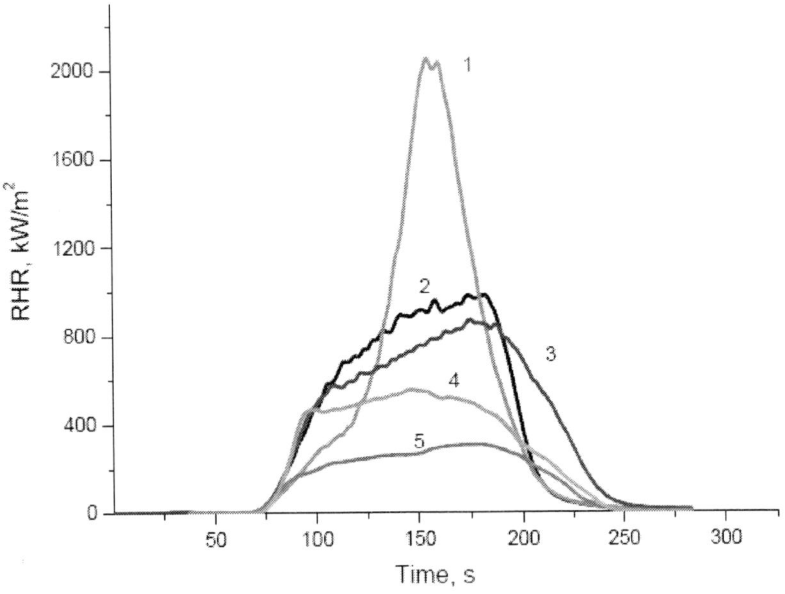

Figure 3. Heat release rate vs. time (incident heat flux of 35 kW/m^2) for:

1. PP,
2. PP–Cloisite 20A (7% wt.),
3. PP–MPP– Cloisite 20A (7% wt.),
4. PP–MPP-Cloisite 20A (5.6% wt.) - MH (20% wt.),
5. PP-MH (50% wt.)

From Figure 3, it is seen that the maximum heat release rate for PP is 2060 kW/m^2, whereas that for the intercalated PP–MMT nanocomposite and the hybrid PP–MPP– Cloisite 20A nanocomposite is 990 and 844 kW/m^2, respectively; thus, the peak heat release rate decreases by a factor of greater than two. It was shown that the time dependence for the specific heat of combustion is practically identical for all of the three samples. This resemblance confirms that the silicate agent does not inhibit the gas-phase combustion process and, thus, does not affect the heat of combustion of the samples, the effect of twofold reduction in the maximum rate of heat release may be explained in terms of formation of a protective char layer on the burning polymer surface. The flammability characteristics of PP with MH (50% wt.) and the nanocomposition of PP–MPP-Cloisite 20A (5.6% wt.) with MH (20% wt.) were also studied with a cone calorimeter at an external heat flux of 35 kW/m^2 (Figure 3). It is seen that that the maximum heat release rate for PP with MH (50% wt.) is 310 kW/m^2, whereas that for the PP–MPP-Cloisite 20A (5.6% wt.) with MH (20% wt.) is 520 kW/m^2.

The peaks heat release rate of PP/ MH (20% wt.) and PP–MPP-Cloisite 20A (5.6% wt.) with MH (20% wt.) decrease by a factor of six and four, respectively, in comparison with pristine PP. It means that relatively small incorporations of MH to PP nanocomposites can substantially reduce the important factor of combustibility, such as maximum rate of heat release.

The main purpose of the present work was the compare the flame retardancy of PP–1, PP compositions with MH (50% wt.) - 5 and composition of PP-PPMA nanofilled by Closite 20A (5.6% wt.) PP with MH (20% wt.) – 4 in order to obtain better flame resistance and to optimize the loading of MH. The composition 4 contains less amount of MH according to the composition 5, therefore it should not lead to dramatic decrease of mechanical properties of PP.

REFERENCES

[1] Wall L.A., *J. Polym. Sci.*, 17, 141 (1955).

[2] Grassie N., *Chemistry of Vinyl Polymer Degradation*, Butterworth, London (1956).

[3] Moiseev Y.M., Neiman M.V., *Vysokomolec. Soed.* A3, 1383 (1961).

[4] Murty MVS, Rangarajan P, Grulke EA, Bhattacharyya D, *Fuel Processing and Technology*, 49:75 (1996).

[5] Ballice L, Reimert R, *Chemical Engineering and Processing*, 41:289 (2002).

[6] Bockhorn H, A. Hornung A, Hornung U, Schawaller D, *Journal of Analytical and Applied Pyrolysis*, 48(2):93 (1999).

[7] Lomakin SM, Zaikov GE, *in Modern Polymer Flame Retardancy*, VSP Int. Sci. Publ. Utrecht, Boston, p. 272 (2003).

[8] Shehata A.B. *Polym Degrad Stab* 85:(2004) 577-582.

[9] Gilman JW, *Applied Clay Sci*, 15:31 (1999).

[10] Gilman GW, Jackson C L., Morgan A B, Harris R H, Manias E, Giannelis E P, Wuthenow M, Hilton D, Phillips S, *Chem. Mater*, 12:1866 (2000).

[11] Davis R.D, Gilman JW, VanderHart DL. Polym Degrad Stab 2003;79:111–21.20.

[12] T. Kashiwagi, R.H. Harris Jr, Xin Zhang, R.M. Briber, B.H. Cipriano, S. R. Raghavan, W. H. Awad, J. R. Shields, *Polymer* 45, 881–891 (2004).

[13] Lomakin S.M., Dubnikova I.L., Berezina S.M., Zaikov G.E., *Polymer Science Series A*, 48 (1) 72 - 84 (2006).

In: New Trends in Biochemical Physics Research
Editors: S. D. Varfolomeev et al., pp. 89-100
ISBN 1-60021-463-0

Chapter 10

Radiolytical Properties of Chitosan and Radioprotection of Cells

O. A.Pilipchatina and V. A. Sharpatyi*
N. M. Emanuel's Institute of Biochemical Physics,
Russian Academy of Sciences
4 Kosygin str., Moscow 111999, Russia

Abstract

In this review the analysis of the experimental data on the radiolytic properties of chitosan and of the modelling its fragments substances is given. It is shown that the main process of radiation chitosan degradation is connected with the formation and transformation of free radicals. The mechanisms of the primary radicals transformation (from initial stage to the final one) up to the formation of some end products of chitosan radiolysis are offered. Problem of the DNA and membrane protection in irradiated cell is discussed.

Keywords: Chitosan; radiolysis; free radicals, scavengers; radioprotection of cell

Introduction

Chitosan is high-polymer of glucosamine. Chitosan of molecular mass from 10 up to 70 kDa is an effective radioprotector: the intravenous injection of it into mouses for 15-30 min befor the irradiation increases the animals survivability up to 73 % [1]. Chitosan is produced enzymically or by the concentrated alkaki treatment from the chitin – the natural analog of cellulo-se, consists of the N-acetyl-α-D-glucosamine residues [2]. The application chitosan as a radioprotector in comparison with others is more preferable because of biocompatibility and

* E-mail: vashar@sky.chph.ras.ru

biodegradation up to glucosamine or N- acetylglucosamine (in the not completely desacetylic drugs) [3] – metaboli-tes of an organism animal and man.

On Free-Radical Mechanism of Chitosan Radiolysis

Despite of more than 40-year's history of radiobiologic investigations on chitosan [1] and available data on it radiolytic properties, generalized in [4,5], complete pattern about the mechanism of a protective action of this polymetric compound, as well as about the mechanism of its radiation destruction, still is not present. The literary data of last years, including investigations on radiolytic properties of polysaccharides and low-molecular carbohydrates simulative the chitosan and chitin fragments, allow us to present the basic directions of their radiation destruction on free-radical mechanisms – sequence of the elementary acts of the primary macroradicals transformations down to formation of some molecular products of radiolysis.

In the present review the results of the radiolytic properties study of the substanses – chitosan precoursors and substanses modelling its separate fragments as of the polymetr compound, and transformations of free radicals in these systems are analysed.

Radiolysis of glucosamine (GA). A attribute reference of the GA radio-lytic properties in water solutions distinguishing from a radiolysis of neutral carbohydrates, is the interaction it with a hydrated electron: $k_{e+GA} = 4.2x10^8$ l mole^{-1} c^{-1}. The participation of the electron is demonstrated by the results of ESR-measuring of the atom H yield in irradiated at 77 K phosphate glass-like GA solutions of different concentrations. The decrease of H yield with elevation of GA concentration reflects passing competing processes:

$$e + H_2PO_4^- = H + HPO_4^= \quad (1)$$

$$e + \text{Glucosamine} = \text{Products} \quad (2)$$

Radicals of glucosamine. Under the OH and H action on a GA as well as in case of neutral carbohydrates the removal of atom H from all of six atoms C is possible The glucosamine residues in the chitosan and chitin mo-lecules are present in the cyclic form, which is characteristic as well as for GA-hydrochloride (GA.HCl), and N-acetylized glucosamine (N-Ac-GA).

At the low-temperature irradiation of the GA.HCl the radicals are formed (with a prevailing yield), ESR spectrum of which is a doublet with *a* ~ 2,6 mT (R_d), caused by interaction of unpaired electron with the proton on C2. At heating of the sample approximately up to 170 K in a spectrum a new wide (ΔH ~ 8 mT) line with five equidistant (~ 2 mT) components arises, interrelation of which intensities, is close to binomial, – quintet (R_q). The change of intensity of the doublet and quintet components, reflecting concentration of R_d and R_q radicals, follows antibate dependence, and the summary concentration of radicals in intermediate temperature range is maintained by a stationary value in limits of errors of measurings. From here it was made the deduction about the transformation of radicals R_d into R_q through the isomerization of R_d with the opening of the residue cycle by break of C5–O5 bond: the atom C5 has an unique place in the radical, where unpaired electron can interreact with the four equivalent protons.

The HFS of the irradiated N-Ac-GA spectrum differs from the HFS of GA and GA.HCl spectra, that is caused by the appearance of the substituent in a N-position and by the formation of radicals in it. The feature of transformation of radicals in N-Ac-GA at the annealing is the formation of allyl-type radicals such as (–C(=O)–ĊH–). The parameters of this radical spectrum (doublet with $a \sim 2$ mT, g=2,007) are caused by a delocalization of a unpaired electron by the fragment of carbonyl group. The formation of allyl-type radicals, as well as at the irradiating of neutral carbohydrates is connected with the primary hydroxyalkyl radicals transformation owing to the H_2O elimination.

Radiolysis of acetamid, N-substituent, simulative a fragment, in chitin. In irradiated at 77 K acetamid ($H_2N–CO–CH_3$) the anion - radical (electron attacks > C=O group; a doublet 1:1, $a \sim 2{,}9$ mT, g=2,003). Under activity of γ- radiation it decomposes with the acylic radical formation ($\Delta H = 0{,}4$ mT, g = 2,0009) [6].

So, from the ESR-measurements on chitosan and chitin low-molecular models follows, that the radicals transformation in the N-replaced unhydroglucose cycle can result in to breakages in a polymer chain and to destruction of monomer units. The participation of an electron is reduced to responses with amino-groups and N-substituents.

Radiolytic properties of heparin. Heparin, a sulfated glucosaminoglu-can, polymer compound of glucosamine and uronic acid residues, connected with glycoside bonds; polyanion; sulfo- and carboxygroups in heparin are the charge carriers (chitosan is polycation). The drugs of heparin with 20-100 monomer unities are applied in medicine as an anticoagulant of blood. By sulphonic-acid groups of a pentasaccharide fragment with a particular sequence of residues (N-acetyl-α-D-glucosamine-6-sulfate; β-D-glucuronic acid; α-D-glucosamine-N-sulfate-3,6-disulfate; α-L-iduronic acid-2-sulfate; α-D-glucosamine-N-sulfate-6-sulfate). Heparin contacts to plusly charged Lis and Arg residues in antithrombin III, and an Arg residue contacts with the second blood plasma protein Xa (coagulative factor Xa). The ternary complex is frameed, in which pentasaccharide inhibits process of formation of a fibrin, i.e. interferes with coagulation of blood [7].

Under the irradiation heparin in water solutions and in a dry view loses anticoagulative properties [8, 9]. In irradiated solutions of heparin hexauro-nic and glucosamine residues are destroyed, the content of nitrogen, sulpho-nic-acid groups is reduced (desulfatation), the molecular mass decreases (depolymerization), the carbo-grours amount in polymer grows and acidity of solution grows as well [8]. The destruction of heparin in solutions, the formation of carboxyl-groups in it descends under influence of OH-radicals ($k_{OH+heparin} = 4.7x10^8$ l mole^{-1}c^{-1}); the e_{aq} reactivity to heparin is less on an order of magnitude [10]. e_{aq}. participates in process of heparin depolymeri-zation, but it, in opinion of the authors [10], is not accompanied by losses of sulphonic-acid groups in polymer.

In heparin water solutions (8-30 %), irradiated at 77 K the radicals of heparin, and of water (OH and $e_{st.}$) are generated, G (e_{st}) ~ 0.1[11]. This value is much more less than one apparent in the 30 % -s' solution of D-glu-cose irradiated in the same conditions: G (e_{st})=0.9. Hence it follows, that at least 90 % of electrons at 77 K react with N-substitutes in glucosamine (in-clouding heteroatoms and olefinic linkages), just as it takes places at a low-temperature irradiation of amino-sugars, peptides and proteins [6]. With the count of the data [8] we concluded also, that the processes of desulfatation and the decreasing of the nitrogen amount in heparin are initiated by *e*.

ESR-spectra of heparin solutions at 77 K are a result of superimposition of basic doublet (1:1) and triplet (1:2:1), centered in region g = 2.003 [10]. As well as in case of carbohydrates [6,12], doublet is from the radical $\dot{C}1$ (splitting on proton). The triplet (splitting on two protons) can belong to the radicals of a removal of H from C2, C3 and C4. The similar spectra (doublet + triplet) in irradiated heparin were observed also in [9,13]. The HFS spectra constants are conditioned by the interaction of an unpaired electron with protons: a = 2.05 mT and a = 2.95 mT [13]. At the heating of irradiated samples up to T = 323 K these radicals are disappeared and only the ions-radicals SO_3^- are registered (the broad single line [9]). The appearance of heparin ions-radicals at 77 K shows the participation of electrons in desusfatation process.

In the case of irradiation of the solid state complexes of heparin –deter-gents (of various composition: cetyl pyridinium chloride, CPC, cetyl dime-thylbenzylammonium bromide and cetyl trimethylammonium bromide) the effective protection of heparin from destruction by the aromatic moieties of complexes is found [14]. So, in the heparin-CPC sample, irradiated at ambi-ent temperature, the radical yield reaches G = 0.2, the value is more than 10 times smaller in comparison with registered one in the components of the complex irradiated individually. ESR-spectrum of the irradiated at 300 K complex heparin-CPC belongs to H-adduct radical of the pyrimidine moiety.

In these conditions of experiment the formation of carboacids and aldehydes, inorganic sulfates, decrease of molecular mass of heparin are not filed. The numbered experimental facts testify to effective protection of heparin in a complex by its aromatic fragment, which can be explained in terms of the free-radical mechanism of energy transfer from heparin to an aromatic moiety of the complex.

Being returned to a main parameter of a functional biological potency of heparin, its anticoagulative properties (parameter showing of integrity and molecular, and supermolecular structures of a polymer compound), we shall note, that filed at an irradiation of heparin the losses of anticoagulative properties can be connected, in our opinion, with 3 factors activity. The main of them is losses of heparin sulphate groups (under influence of an electron) and from here the impossibility of formation of a ternary complex "pentasaccharide - antithrombin III -Xa ". The 2-nd factor is the change of a conformation of heparin chain because of formation of epimer in any of the pentasaccharide residues (the conformity, complement of a locating of heparin anions and plusly charged amino-acid residues is broken). The epimers appearance process on response $R + e = R^-$ in the macromolecules with carbohydrate moieties (for example, in DNA) surveyed in [15]. The3-rd factor is the transformation of any primary radical in pentasaccharide with the breakage appearance in a macromolecule, that also interferes with formation of the specified complex.

Chitosan Radiation Chemistry

The main processes of chitosan destruction are: deamination (G ~ 3), formation of breaks in polymer chain (G = 4.2 ± 0.3), abjection of molecular hydrogen (G = 1,3) and dioxide of carbon (G = 1,5) [4]. In irradiated at 77 K chitosan and chitin ESR method registered the radicals. The sum yield of radicals at 77 K in chitosan is 3.0, in chitin $G(\Sigma R) = 2.3$.

In case of the chitosan irradiation at 77 K, in which ~ 30 % of amino-groups are replaced by the quaternary ammonium groups (chloride ammoni-um), the radicals yield is G = 1.9, the of anions-radicals Cl_2^- yield is G = 0.04. At the chitosan and trioxybenzoic acid (3,1 % of masses.) conjugate irradiation at the same conditions the yield of radicals decreases up to 0.76, and the yield of ions – up to G = 0 [16]. These data specify that, as well as in surveyed earlier system "heparin – aromatic compound (AR)", at the com-plexes chitosan-AR irradiation the transfer of energy, immersed by a poly-mer, the second moiety of this complex also is effective. Hence it appears, that if heparin and chitosan are used in radiation medicine, that for the radi-oprotection of both polymers it is necessary to bring in their composition "small quantities" of the “AR”.

Mechanism of chitosan radiolysis. The main difference of aminosugars radiolysis from radiolysis of neutral ones is the fact that they (aminogroups) react with electron. The result of the interaction of **e** with aminogroups of chitosan ($k_{e+chitosan}$ = 4.6x10^8 l.mole^{-1}c^{-1} [5]) is the deaminization of the poly-mer [39]: alkyl radicals C2 and ammonia are formed. The similar processes are revealed in proteins, irradiated at low temperature, – deamination of Lis aminoacids residues [6]. In irradiated at T = 300 K chitosan carboxyl- (G ~ 0.6) and carbonyl groups are accumulated, and the yield of reducing groups is 25 ± 2 (it was determined by copper number) [4, 17,18]. The mechanisms of these radiolytic products formation are analogous to the ones examined recently for the neutral polysaccharides radiolysis [6].

The free-radicals structure and transformations of them. By ESR spect-ra in irradiated at low temperature chitosan and chitin the the primary radi-cals of H removal from C1 (doublet 1:1, *a* ~ 3,0 mT) and from C4 (triplet 1:2:1, *a* ~ 2,6 mT) were identified. Under the heating of the irradiated chitin (acetylated chitosan) up to T ~ 200 K the concentrations of $\dot{C}$1 and (essen-tially) $\dot{C}$4 were decreased and the spectrum of radical GluNH$\dot{C}$(OH)CH_3 (1:3:3:1, *a* ~ 1.4 mT) was registered, where Glu was the fragment of poly-mer chain [17, 18]. Analysis of presented in [17, 18] ESR spectra of irradia-ted chitosan and chitin allows us to determine supplementary the line with the total width ΔH ~ 9,5 mT registered at 77 K against a background of others (and for some reason not taken into account by the authors) and to in-terpret it as doublet of triplets(a_d~ 2 mT, a_t ~ 3,8 mT, ΔH_t~ 1 mT,g ~2,003) attributing it to the radical of the aminogroup from C2 removal. Concentra-tion of these radicals at 77K is 20-25 % from total amount of radicals [19].

The authors of work [5], studying the chitosan radiolysis in aqueous solutions (pH~ 3, saturated by N_2O, the radicals OH are the reacting agents), specify an opportunity of the formation of radicals of the H removal from all atoms C in the unhydrocarbohydrate cycle and explain the breaks in a poly-mer chain formation by the transformations of the primary hydroxyalcylic radicals C1, C4 and C5 [20,21]. For the radicals C1 and C4 two variants of such transformations are supposed: a hydrolysis of glycoside bonds in the radicals C1 and C4 at the β-position and the radical C1 fragmentation at the β-bond position with the lactone formation and transmission of unpaired electron on the next monomer unit. The transformation of radical C5 accom-panied by disclosing of unhydrocarbohydrate cycle with appearance in it carbonyl groups on C5 and C1 and the jumping of unpaired electron on C4 of the next monomer unit. At the presence O_2 α-hydroxy-peroxy-radicals C2, C3 and C6 are arised, transformation of which accompanies by the elimina-ting HO_2 with the carbogroups formation on these atoms C. It is underlined, that aminogroups of chitosan during the radicals transformations behave similarly to OH-groups [5].

Radicals C1 may transform by two directions - isomerization of Ċ1 with break of C5-O5 and carry-out of unpaired electron on C5. Such process is registered by ESR method at low-temperature irradiation of glucosamine [22]. Further stages of transformations of radicals C5 in the presence of O_2 for example in dry starches lead to formation of malic or tartaric acids [21].

The other one direction of the Ċ1 transformations is isomerization with break of a glycoside bond Ċ1-O1-C'4 and formation of keto-group at C1 (lactone) with unpaired electron carry-out to C'4 of neighbouring monomer unit (as well as is supposed in [5]). With the transformations of a radical Ċ'4 at presence O_2, the accumulation of compounds of 2-amido-2-desoxytetradi-aldose (from glucosamine residue) and 2-acetamidodesoxytetra-dialdose (from N-acetylsubstituted residue of glucosamine) type is connected. The last product was obtained under irradiation at presence O_2 saturated N_2O solutions of N-acetyl-2-glucosamine (G = 0.5, 0.01 M [12]).

At the presence of water in the radicals Ċ1-O1-C'4 the bond O1- C'4 can be hydrolyzed [5,23], and at the atoms C1 and C'4 in the next monomer unit the hydroxyl groups are appeared. During the transition of the H from hydroxyl C1OH to O5 an acyl-radical is formed (on C1), eliminating CO; CO in chitosan was registered [12,17].

For radical C2 three variants of transformations are possible [19]:

(a) by the type of water β-elimination (removal of H from C3 and OR from C4). The break appears in polymer chain and the desoxy (C4) - and keto (C3) -groups are formed in monomer unit at the same places as under citric and siccine acids formation in starches [21];

(b) isomerization with break of bond C3-C4; by the formation of double bond C2=C3 and radical with non-coupled electron at C4 under radiolysis of glucosamine in solutions the glycerin aldehyde was registered (in 0,05M solution G = 1,3 [12]);

(c) regrouping of bonds H–C at C1 and C2 with carry-out of non-coupled electron on C1 and further transformation of radical C1 analogously to radical C2 transformation in the case of cellulose [19,20]. In subsequent ones the compounds are presented identified under radiolysis of glucosamine or N-acetylglucosamine [12].

The proof of an opportunity of embodying of variant "a" is served by detection among products of the starch radiolysis of the citric and succinic acids keeping at atoms C3 and C4 the specified functional groups. The sequence of the elementary acts of this process in details surveyed in [20, 21].

The transformation of the alkyl radical C2 on a route "b" would prove to be confirmed by findings of investigation of a radiolysis of glucosamine in solutions (0.05 M). In this case glyceraldehyde is registered, G = 1.3 [12].

The first stage of the radical C2 transformations on a route "c" is redistribution of the H-C bonds at atoms C1 and C2 in a cycle, that is similar to that descends in case of cellulose [19]. The radicals C1 can transmute on two directions - isomerization with a breakage of C5–O5 bond and run-over of a unpaired electron on C5. Such process is filed by a method ESR at a low-temperature irradiating of glucosamine [6,22]. The subsequent stages of C5 radical transformation, for example, in irradiated in presence O_2 dry starches give in malic and tartaric acids [21].

If under chitosan irradiation in the presence of O_2 the desamination act is accompanied by formation of peroxide radical C2OO and then by formation of peroxide (by C2) then under its decomposition with break of bonds O–O and C1–C2 (as well as in irradiated starches [24]) the aldehyde group is appeared (C2). As a result of regrouping of bonds at C1 the arabinose may be formed.

If the sample of chitosan contains besides of glucosamine residues the N-acetyl-β-D-glucosamine (not fully deacitylated chitin moieties),we should take into account an opportunity of electron capture by carbonyl group of acetyl with formation of anion - radical. The similar process is observed in case of the low temperature irradiation of acetamide, of proteins and pepti-des: electron attacks carbonyl group of peptide bond. Identifiable at 77 K in the irradiated proteins anion-radicals (poor-resolved doublet with $a \sim 1.5$ mT) at the sample heating up to 190 – 200 K is protonized (at interaction with molecule of water finding mobility owing to structural change). The neutral radical - H-adduct with the approximately same ESR-parameters is formed, as well as in the initial anion- radical (poor-resolved doublet R_p [6]).

Parameters of the ESR-spectrum and conditions of chitin irradiattion and the GluNHĊ(OH)CH_3-radcial registration [4,17] allowed concluding that in given system the H-adduct of acetyl from anion-radical is formed. Hydrolysis of N-C in this radical is accompanied by the formation (with equal yields) of free amino-group and molecule of acetic acid [4]. Such interpretation of effect of deacetylation in chitin [4,17] is seemed for us more acceptable.

The schemes reflect the participation of radicals appearing under the action of electrons. With consideration of the values of yield of chitosan de-struction process (formation of breaks) and deamination yield we may assu-me that more than 2/3 of effect (71 %) is fallen at the last one, and consequ-ently at the participation of electrons in polymer destruction. The rest 29 % reflect the contribution into destruction of primary radicals of removal of H.

Investigations of radiation chemistry of glucosamine and N-acetylglucosamine were carried out at solutions concentrations 0,01-0,1M, i.e. at conditions of total involving of forming from water radicals into reactions with dissolved substances [12]. Electrons attack the fragment N-C2, and radicals OH attack H–C bonds of all 6 carbon atoms, radicals of H removal are formed [18]. The last fact is confirmed by the structure of identified products of radiolysis in the absence of O_2, i.e. by compounds containing keto- (or aldehyde) group at each of carbon atoms. Mechanisms of formation of these compounds are considered in [12].

The schemes (1, 2) are given on the basis of the dry polysaccharides ra-diolysis examinations. They are applicable and for a liquid-phase radiolysis, as and in this case aminogroup leaves under influence of an electron.

Analysis of a radiation chemistry of glucosamine and N-acetylglucos-amine in aqueous solutions were carried out at concentrations 0.01–0.1M,i.e. in the conditions of complete recruitment phenomenon of radicals, gene-rated from water, in response with solutes (the effect of an indirect action of radition is peak) [12]. The electrons attack a fragment N–C2, and the OH at-tack the bonds H–C of all 6 atoms C, the radicals of H removal are formed [6]. The last was proved by the composition of identified products of radio-lysis (in absence of O_2) – substanses keeping keto- (or aldehydic) groups at each atom C. The mechanisms of formation of these bonds surveyed in [12].

Under irradiation of dry chitosan the value of yield is $G(H_2) = 1.3$, and. it may be used as basic one under estimation of contribution of radicals of H removal into total process of polymer destruction, if we remember that H_2 in polysaccharides is formed by the reaction ·H + HR = H_2 + ·R [17]. Then the part of H removal radicals in chitosan destruction is 1.3 / 4.2 ~ 0,3. so, we get the same value equal to 30 %, as well as that one determined above by the difference between the yield of sum destruction process and contribution into this process of electrons conditioning desamination of polymer. Direct measurement of radicals concentration in chitosan at low temperature gave the value G = 3.0 [4]. This value exceeds the estimation one, but it is lower if we assume, that one radical causes one break in a polymer chain.

Formation of CO_2 in irradiated chitosan (or chitin), keto- and desoxy-ketoderivatives of glucoamine in solutions of chitosan, glucosamine and N-acetylglucosamine is connected with transformations of radicals of H remo-val. The scheme of the mechanism of CO_2 formation in the absence of O_2 (that means that by C1, since only it is connected with oxygen atoms [23]).

Formation of arabinose is connected with transformations of radical C1 in the absence of O_2. Mechanism of this process for solutions of glucosami-ne and N-acetyl-substituted glucosamine is considered in [12]: after hydrol-ysis in ·C1 of bond β -O1–C'4 and transition of H on O5 the acyl radical C1 eliminated CO, and in the case of N-acetylsubstituted glucosamine the radi-cal –Ċ2HNHAc is formed. Under its interaction with water acetamide is removed and end-hydroxyradical is formed, oxidation of which (removal of H from hydroxyl) leads to formation of arabinose G (arabinose) = 2.1.

Formation of keto- and desoxyketosugars of N-acetylglucosamine [12] with consideration of data on cellulose and starches [21,24] represents multi-stage process of transformations of the H removal primary radicals. Trans-formations, for example of ·C3 include the following stages: " β-elimination of water " with participation of cyclic atom O [28], regrouping of bonds in the allyl type radical, its isomerization with bonds C1–C2 and O–H breaks in hydroxyl at C1 and removal of CHO. the last one is dimerized with formation of glyoxal registered under cellulose and starches irradiation.

Process of formation of desoxyketogroup –C4H_2–C3 (=O) – in cycle of N-substituted glucosamine is presented in scheme 4b [24]; in polymer the hydrolysis of β-bond in glycoside radical C1 and break of C1–C2 bond with carrying-out of H from hydroxyl C1 on atom C2 occur. Further hydrolysis of bond C5–O5 in radical leads to 1-acetamido-1,3-didesoxypent-2-ulose regis-tered in irradiated 0,01M solution of N-acetylglucosamine (G = 0.1 [12]).

In conclusion we should note two moments:

- 30 % of effect of chitosan destruction in the absence of O_2 fall at the part of primary radicals of removal of H, and the value of yield of only one product CO_2 equal to 1.5 even slightly exceeds these 30 %. From the other hand, the yield of radicals in chitosan is lower than the destruction yield. We may suggest that some of radicals while transforming transmits unpaired electron to neighbor monomer units and initiates their decomposition. For example, in the case of cellulose it was found that as a result of radical C1 transformation 2 neighbor monomer units are decomposed due to transfer of non-coupled electron from C1 to C'4. In the upshot we obtain that without change of sum amount of radicals in system the amount of products of their transformations is increased. Mentioned value of yield of formation of redu-cing (carbonyl) groups in irradiated chitosan (~ 8 carbonyl groups per 1 radi-cal) serve as confirmation of the last fact. Destruction of neighbor monomer cycles doesn't influence on number of breaks in polymer;

- chitosan destruction is caused by participation of both the electron and radicals OH and H. It means that chitosan under irradiation of aqueous solutions may effectively capture the water's radicals[25]. According to [26], chitosan is effective scavenger of alkyl, hydroxyl and superoxide radicals. Considering the structure of linear polymer, the presence of two hydroxyl and one amino groups in monomer unit which may form in water molecule hydrogen bonds as it is proved in the case of starch (structuring of water by it) we want to note that chitosan obviously possesses the same properties. Its high activity as radioprotector is probably caused by ability of polymer to structure the water (gelatinization) and consequently to capture water radi-cals " at distance " [27]: at intravenous introduction of chitosan with mole-cular mass from 10 up to 70 kDa to mouses from calculation 20mg/kg for 15-30 min before the irradiation in dose 8 Gy (minimum absolutely lethal) increases the animals survivability up to 73 % [1]. This fact may serve as an illustration of radioprotection by mechanism of competition for radicals of water, and revealing of so-called effect of indirect irradiations action.

On Mechanisms of Cell Protection by Chitosan

Being returned to a problem of radioprotection of a cell, organism by chitosan and taking into acount of embodying of the indirect action mecha-nism of ionizing radiations, we shall make attempt to present chitosan as DNA and membranes (two basic cell targets) protector [28].

Chitosan in a role of DNA protector. By experience in vitro [29,30] shown, that DNA and chitosan can form complexes.Chitosan-DNA nanopar-ticles of the size 100-250 nm and of the composition 35.6 weight % DNA and 64.4 % chitosan are formed owing to electrostatic interaction between amino groups of ligand and phosphate groups at ratio N/P = 3 - 8 (at chito-san concentration of 100 microg/ml) [31]. The formation of similar particles with a ratio N/P = 4 and the size of 400-600 microns (200 microg/ml) is re-vealed and on the other system - at preparation of an anticaries DNA vaccine (plasmid pGJA-P/ VAX) in chitosan [32]. The transfection efficacy depends on a type of used cells. The opportunity of chitosan penetration through cy-toplasmic membrane of various cells is scored also in [33]. If chitosan insi-nuates into a cell and forms with DNA a complex, as in vitro [29,30], being in a composition of complex it could represent itself as the free radicals sca-venger,similarly to that descends in a case of sulfurcontaining amines– radi-oprotectors [34]. (The latter ones are attached with the amino groups to ne-gative charged atom of oxygen in DNA phosphates[35].) Opportunity of formation of complexes"sulfurcontaining amine – DNA" and scavenging of free radicals is confirmed by apparent decrease of a mutation frequency in irradiated animals: mouse, rat, rabbit (injected with the radioprotectors) and Drosophila (radioprotector diet) [36].

It is possible to figure, that the basic mechanism of chitosan penetration in a cell as well as for other macromolecules (proteins, lipids, glycoproteins) is endocytosis. The sizes of chitosan macromolecules in a drug of maximal efficiency (at MM = 10-70 kDa [1, 37]) are compounded with the typical ve-sicle dimensions (from 50 up to 1000 nm) for polymer molecule at endocy-tosis. The times of chitosan drugs introduction into animal before the irradia-tion for the achievement of maximal effect of radioprotection [1] correlate to the time of the specified macromolecules penetration into a cell under endo-cytosis and of a ligand–receptor associates formation (up to10 min [38]).

On chitosan radioprotection of a membrane. If, as well as in a case of DNA, we talk about avoidance of free radicals action on the main frames of a membrane, it is necessary to take into account two factors. The first one: to what membrane fragments the chitosan molecule attaches; the second one: chitosan protects membrane from what types of damages.

Practically all the cytoplasmic membranes of animal cell contain phos-pholipides with negatively charged heads - phosphatic groups, from here it is necessary to figure, that the chitosan protonated aminogroups could be asso-ciated with them. The degree of interlinking them with membrane phospha-tic groups depends on pH of solution, and on molecular mass of chitosan. The interaction of chitosan with lipids of a membrane is testified by the fact, that in this place exactly the order of lipid bilayer organization is variated [39]. At a low-temperature irradiation of aqueous solutions of lyposomes dispersions the radicals of lipids are formed; the yield of them is essentially reduced at the presence of a radioprotector (sterically hindered phenol, ac-ceptor of free radicals), introduced into lipid bilayer of lyposomes before the irradiation [40]. These data have allowed to assume, that chitosan, being attached to phosphatic heads of lipids of membranes, can represent itself as scavenger of free radicals – water radiolysis products.

Proteins are the second main component of membranes. In dependence on a type of cell the content of them in membranes changes from 25 % up to 75 %. The basic function of membrane integral proteins (interesting for us) is to carry out the transport of matters into cell, which are necessary for its functioning.The system of receptors of lipoprotein and glycoproteins – sub-units in a composition of integral protein checks this process of transport. Under the irradiation the structure and functioning of integral proteins are broken, and through a defective membrane any antigens and toxines can pe-netrate into the cell. By experiments [41] on macrophages of bat cells (RAW 264.7) and others ones [42] was shown, that the chitosan aminoglucose resi-dues were associated with receptors of immunoglobulins specific to manno-se, on a system of the complement (key - lock). From here it is possible to conclude, that chitosan interacting with receptors of the immunoglobulin (the protein fragment of a glycoprotein complex) interferes attack by free radicals and maintains the receptor system for the subsequent functioning - recognition of antigens. In this case we deal with the conservation of immunity of an organism supressed, as is known, by the radiation.

Thus, the presence of chitosan on a membrane can prevent (or reduce) the changes of structural organization of a bilayer of lipids and the destruction of receptors under action of ionizing radiations.

REFERENCES

[1] Il'in L.A.,. Andrianova I.E, Glushkov V.A. // *Radiation biologiya. Radioekolo-giya,* 2004. V. 44. № 2. C. 176 (in Russian).

[2] Stepanenko B.N. *Chemistry and biochemistry of carbochydrates* (polycuccharides). Moscow : Vysshaya shkola. 1978. 256 p.

[3] Tsygan V.N., Zhogalev K.D., Nikitin V.Yu. // *Rynok BAD*,2002.№2.P.1.

[4] Ershov B.G., *Uspekhi khimii,* 1998. V. 67. № 4. P. 353. (in Russian).

[5] Ulanski P., von Sonntag C. // *J.Chem. Soc., Perkin Trans.*, 2000. V. 2. P. 2022.
[6] Sharpatyi V.A. *Radiation chemistry of biopolymers*. Moscow: Energoizdat.1981. 168 p. (in Russian).
[7] Van Boeckel C.A.A., Petitou M. // *Angew. Chem. Int. Ed. Engl.,* 1993. V. 32.№ 12. P. 1671.
[8] Balazs E.A., Laurent T.C., Howe A.F., Varga L. // *Radiat. Res.*, 1959. V. № 2. P. 149.
[9] Kudryavtsev Val.N., Grushevskaya L.N., Yurik T.K., Kabanov V.Ya.. // *Khimiya vysokikh energii,* 1993. V. 27. № 1. P. 41 (in Russian).
[10] Baugh P.J., Morgan R.E., Kershaw K. Phillips G.O. // *Radiat. Res.*, 1971. V. 46. № 2. P. 217.
[11] Jooyandeh F., Moore J.S., Morgan R.E/, Phillips G.O. // *Radiat. Res.*, 1971. V. 45. № 3. P. 455.
[12] Kochetkov N.K, Kudryashov L.I., Chlenov M.A., *Radiation chemistry of carbohydrates,* Moscow: *Nauka*, 1978 288 p (in Russian).
[13] Baquey Ch, Darnez Ch. // *C.R.Acad. Sc.*, 1973. Ser. C. T. 277. P. 139.
[14] Moore J.S., Phillips G.O., Rhys D.// *Radiat. Res.*, 1972.V. 50. № 3. P. 479.
[15] SharpatyiV.A. // *Radiobiologiya*, 1992. V. 32. Vyp. 2. P.180 (in Russian).
[16] Aleksandrova V.A., Shilova I.A., Kuzina S.I., Mikhailov A.I. // *Proc. Of IV N.S. Bakh conference on radiation chemistry.* Moscow 1-3 June 2005. P.10 (in Russian).
[17] Ershov B.G., Isakova O.V., Rogozhin S.V., Gamzazade A.I., Leonova E.Yu., // *Doklady Akad. Nauk SSSR*, 1987. V. 295. № 5. P.1152 (in Russian).
[18] Ershov B.G., Sukhov N.L., Nud'ga L.A.,. Baklagina Yu.G, Kozhevnikova L.G., Petropavlovskii G.A. // *Zh. Prikl. Khimii,* 1993. V. 66.Vyp.3. P. 649 (in Russian).
[19] Sharpatyi V.A., / *Chemical and biological kinetics. New horizons.* V. 1. Chemical kinetics. Chapter 6. Ed.by: E.B.Burlakova, S.D.Varfolomeev, G.E.Zaikov, A.E.Shilov. Moscow: Chimiya.2005. P. 108.
[20] Pristupa A.I., Sharpatyi V.A. // *Doklady RAN*, 2002. V. 387. V. 387., № 5. P. 643 (in Russian).
[21] Korotchenko K.A., Pristupa A.I., Sharpatyi V.A. // *Khimiya vysokikh energii*, 2004. V. 38. № 2. P.107 (in Russian).
[22] Nadzhafova M.A., Sharpatyi V.A. // *Doklady Akad. Nauk Azerbaidzh.*, 1977. V. 33. № 8. P. 41 (in Russian).
[23] Kvach N.M.,Kuvaldina E.V., Sadova S.F., Sharpatyi V.A. // *Doklady RAN,* 1996. V. 349. № 1. P. 60 (in Russian).
[24] Sharpatyi V.A., Shapilov A.A., Pintelin S.N. // *Khimicheskaya Fizika,* 2001. V. 20. № 12. P.19 (in Russian).
[25] Nishimura Yo., Kim Hee-S., Ikota N., Arima Hi., Bom Hee-S.,Kim Yo.- Ho, Watanabe Yo.,Yukawa M., Ozawa T.// *J. Radiat.Res*, 2003.№ 1. P.53.
[26] Park Pyo-J.,Je J-Y.,Kim S.-K.//*Carbohydrat Polym*, 2004. V. 55. P.17.
[27] Korotchenko K.A., Sharpatyi V.A. // *Radiation biologiya. Radioekologiya*, 2000. V. 40. № 2. P. 133 (in Russian).
[28] Burlakova E.B., Shishkina L.N. // The problems of natural and modifying radiosensitivity. / Ed. by M.M.Konstantionova and A.M. Kuzin. Moscow: *Nauka*. 1983. P. 29 (in Russian).
[29] Nechipurenko Yu.D., Volf A.M , Salyanov V. I., Evdokimov Yu.M. // *ZhETF*, 2004. V. 125. Vyp. 1. P. 103 (in Russian).

[30] Evdokimov Yu.M., Salyanov V. I., Krylov A.S.,Nechipurenko Yu.D., Volf A.M. // *Biofizika,* 2004. V. 49. Vyp. 5. P. 789 (in Russian).
[31] Mao Hq, Roy K, Troung-Le Vl, Janes Ka, Lin Ky, Wang Y, August Jt, Leong Kw // *J. Control Release,* 2001. V. 70. № 3. P. 399.
[32] Li Yh, Fan Mw., Bian Z., Chen Z., Zhang Q., Yang Hr. // *Chin. Med. J.* (Engl.), 2005. V. 118. № 11. P. 936.
[33] Prabaharan M., Mano Jf // *Drug Deliv.*, 2005. V. 12. № 1. P. 41.
[34] Eidus L.Kh. *Phisiko-chemical principles of radiobiological processes and protection from irradiation.* Moscow: Atomizdat. 1979. 216 p (in Russian).
[35] Kuzurman P.A., Sharpatyi V.A. // *Radiation biologiya. Radioekologiya,* 2006. V. 46. № 1. P. 22 (in Russian).
[36] Mosse I.B. *Radiation and heredity: genetic aspects of radioprotection.* Minsk: Universitetskoe, 1990. 205 p (in Russian).
[37] Mori T., Murakami M., Okumura M., Kadosawa T., Uede T., Fujinaga T. // *J. Vet. Med. Sci.*, 2005. V. 67. № 1. P. 51-56.
[38] Varfolomeev S.D., Gurevich K.G. *Biokinetika.* Moscow: Grand, 1999. 720 p (in Russian.
[39] Fang N., Chan V., Mao Hq., Leong Kw. // *Biomacromolecules,* 2001. V. 2. № 4. P. 1161.
[40] Paramonov D.V., Trofimov V.I., Knyazev A.A. // *Khimiya vysokikh energii,* 2004. V. 38. № 2. P.113 (in Russian).
[41] Han Y., Zhao L., Yu Z., Feng J., Yu Q. // *Int. Immunopharmocol.*, 2005. V. 5. № 10. P. 1533.
[42] Feng J., Zhao L., Yu Q. // *Biochem. Biophys. Res. Commun.*, 2004. V. 317. № 2. P. 414.

In: New Trends in Biochemical Physics Research
Editors: S. D. Varfolomeev et al., pp. 101-109
ISBN 1-60021-463-0

Chapter 11

STATE OF LIPID COMPONENT OF SOYBEAN FLOUR DURING STORAGE

L. N. Shishkina, S. E. Traubenberg*, E. V. Miloradova*, I. V. Vialtseva* and A. A. Kozlova*

N. M. Emanuel Institute of Biochemical Physics,
Russian Academy of Sciences, 4 Kosygin str., Moscow 119991, Russia
*Moscow State University of Food Production,
Volokolamskoe shosse, Moscow 125080, Russia.

ABSTRACT

It was studied the effect of packing on the state of lipid component SOPROPOLETS-8-TB-325 of fat-free and lecithin enriched soybean flour during storage within a period of three months. It was shown a credible change of a large variety of parameters of the lipid component of fat-free and lecithin enriched soybean flour already after one month since the beginning of experiment. The obtained data make it possible to draw a conclusion on preferable use of multi-layer (aluminum) packing rather than mono-layer (transparent) packing for storage of fat-free and lecithin enriched soybean flour.

Keywords: Fat-free and lecithin enriched soybean flour; lipids; stability during storage

AIMS AND BACKGROUND

One of the advantages of powder products is the possibility of their long shelf-life. It was established that lability of lipids causes damage to edible fats in the process of their storage [1, 2] and, therefore, the duration of storage of dry mixture is also affected by keeping of the lipid component of system [3, 4, 5]. There are data on significant change of lipid component of wheat flour [6], dry yeast membranes [7] and doughnuts mixture [5] during storage, which makes it possible to suppose a course of complex hydrolytic and oxidative processes in the lipid component during storage of dry food powders. Owing to the modern technique of soy processing, a soybean flour of new generation and soy-based products produced by the SOYPROTEIN plant (Yugoslavia) [8, 9] are now more widely used as a food additive. One of these products is heat-treated fat-free and lecithin enriched soybean flour SOPROPOLETS-8-TB-325 [9].

Fat-free soy flour contains about 38% of total carbohydrates including 15% of soluble mono- and oligosaccharides, and 13% of polysaccharides [10].

The flour with recombined fat content is obtained by adding lecithin is to it in order to reduce the dust formation, to improve the dispersibility of flour and enhance its nutritional properties. However, the presence of lecithin being a mixture of natural phospholipids [11] can obviously affect the keeping quality of flour. Besides, the state of the lipid component may also depend on the type of packing.

In connection with the aforesaid, the purposes of research was the study of the influence of the type of packing on the state of lipid component of the fat-free lecithin enriched soybean flour during storage.

MATERIALS AND METHODS

The following materials and methods were used:

- heat-treated fat-free soybean flour obtained from selected ecologically pure and genetically non-modified peeled soybeans with addition of lecithin SOPROPOLETS-8-TB-325 with fineness degree of 0.045 mm (the Scientific Institute of agriculture and vegetable farming, Novi Sad, Yugoslavia, 2004) [12].
- lecithin, date of manufacture is 20.01.2005, Specs TC 9145-030-51070597-2003.
- mono-layer (transparent) packing according to Specs TC 9761-001-43518761-01.
- multi-layer (aluminum) packing according to Specs TC 9761-001-43518761-01.

The investigated samples were simultaneously air-tightly packed into different types of packing and kept in dry places at the room temperature. The characteristics of flour measured at the beginning of the experiment served as the reference standards.

The content of secondary products of oxidation was determined by their capability of reacting with thiobarbituric acid (TBA-reactive substances, TBA-RS). The amount of TBA-RS was determined using method [13]. The sample protein content was measured by the modified microbiuret method [14].

Lipids were isolated from the samples using the Blay and Dyer methods in the Kates modification [15]. The isolated lipids were used for studying the composition of phospholipids (PL), the amount of sterols (%) and the contents of peroxides. The amount of peroxides in lipids was determined by the iodometric titration method according to the procedure specified in Russian standards 26593.

The qualitative and quantitative composition of PL was determined by the thin-layer chromatography [16] with the use of silica gel of type G (Sigma, USA) and glass plates measuring 9 ×12 cm. The chloroform – methanol – glacial acetic acid- water mixture in the ratio of 50:30:8:4 [16] was used as the mobile phase. The development of chromatograms was performed by iodine vapour. The quantitative analysis of the PL composition was determined after the removal of fractions from the plate and after PL perchloric acid digestion to inorganic phosphate. For the colour reaction to phosphorus we used ammonium molybdate and ascorbic acid, produced by Serva (Germany) and also perchloric acid of chemically pure grade. The amount of inorganic phosphorus was judged according to the optical density of the solutions at the λ=800 nm wavelength as measured with spectrophotometer KFK-3 (Russia). Plotting of the calibration curve was carried out on the monosubtituted potassium phosphate of specially pure grade. The sterol content in lipids was determined spectrophotometrically at 625 nm wavelength [17]. Serva Company cholesterol was used for plotting the calibration curve.

Among the generalized parameters of lipids composition the contents of total lipids (TL) in one (1) gram of absolutely dry matter (ADM), the amount of PL (%PL) and sterols (%) in the total lipids, phosphatidyl choline/phosphatidyl ethanolamine (PC/PE) ratio, since these fractions are the main in the PL composition of the vegetative objects; molar ratio [sterols]/[PL] and the ratio between the sums of the more easily oxidizable to the more poorly oxidizable PL fractions (∑EOPL/∑POPL) were examined. The latter ratio was calculated by the formula [18]:

$$\sum EOPL/\sum POPL = (PI+PS+PE+PG+CL+PA)/(LPC+SM+PC),$$

where PI is phosphatidyl inositol, PS is phosphatidyl serine, PG is phosphatidylglycerol, CL is cardiolipin, PA is phosphatidic acid, LPC are lysoforms of PL, SM is sphingomyelin. The more easily oxidizable phospholipids are the fractions in the composition of which there are predominantly unsaturated fatty acids and the more poorly oxidizable phospholipids are the fractions in the composition of which there are predominantly saturated fatty acids in the hydrocarbon part molecules. That ratio makes it possible to judge about the oxidizability of lipids [18].

The determination of the contents of lipids, the amount of peroxides in lipids, the contents of TBA-reactive substances in the investigated objects was conducted in three independent samples, repeating each measurement from two to five times. The experiment data were processed with a commonly used variational statistic method [19]. The experimental data are presented in the tables and figures in form of arithmetic means with the indication of the mean square errors of the arithmetic mean (M ± m).

Results and Discussion

During the first stage the composition of phospholipids of the "lecithin" product was analyzed. The results obtained show that lecithin is actually a mixture of natural lipids among which 21.0 ±1.0 % are represented by phospholipids and one third of the total composition of phospholipids is attributed to phosphatidyl choline (Table 1). The generalized parameters are as follows: PC/PE = 1.384±0.043, and ∑EOPL/∑POPL = 0.953±0.048. The composition of phospholipids of the initial soybean flour is presented in Tables 2, 3 (column "0").

The comparative analysis of the relative contents of PL fractions and the generalized parameters of the lipids composition of the "lecithin" product and fat-free lecithin enriched soybean flour shows the absence of significant differences in the relative contents of the PL main fractions (PC and PE) and their ratio. This makes it possible to assume that the soybean flour lipids are predominantly the lipids of lecithin added to it.

Table 1. The composition of phospholipids of the "lecithin" product

Fraction, % P	n*=5
LPC	6.4±0.45
SM	12.05±0.65
PC	32.85±1.75
PI	10.65±0.65
PS	2.80±0.50
PE	23.75±1.05
CL+PA	11.50±2.60

*n – number of parallel measurements.

Table 2. The composition of phospholipids of fat-free and lecithin enriched soybean flour during storage in a multi-layer (aluminum) packing

Fraction, % P	Duration of storage, months			
	0	1	2	3
	n*=5	n=5	n=15	n=14
LPC	4.00±0.45	3.65±0.26	3.10±0.50	4.05±0.40
SM	2.70±1.55	3.15±0.25	2.80±0.25	2.50±0.20
PC	38.7±3.40	42.80±1.65	47.55±1.10	51.50±1.05
PI+PS	18.45±4.25	18.45±1.65	14.55±0.50	11.40±0.60
PE	28.55±3.0	25.55±1.30	23.9±1.10	26.50±0.50
PG	-	1.15±0.50	-	-
CL+PA	7.60±1.45	5.20±1.55	8.10±0.65	4.05±0.40

*n – number of parallel measurements.

Table 3. The composition of phospholipids of fat-free and lecithin enriched soybean flour during storage in a mono-layer (transparent) packing

Fraction, % P	Duration of storage, months			
	0	1	2	3
	n*=5	n=13	n=15	n=14
LPC	4.00±0.45	5.45±1.15	4.45±0.35	6.25±0.65
SM	2.70±1.55	5.80±1.20	4.95±0.45	7.70±0.90
PC	38.70±3.4	39.15±1.25	45.35±1.30	42.50±1.25
PI+PS	18.45±4.25	17.65±1.30	16.4±1.30	10.70±1.90
PE	28.55±3.0	22.85±1.70	24.15±0.65	24.10±0.75
PG	-	1.50±0.40	-	-
CL+PA	7.60±1.45	7.65±1.10	4.75±0.55	8.75±1.00

*n – number of parallel measurements.

The next stage of investigations was the study of the state of the flour lipid component during three months of storage in two types of packing. The PL composition of examined samples is presented in Tables 2 and 3. The dynamics of parameters characterizing the contents of oxidation products and amount of total lipids in one (1) gram of ADM is presented in Table 4.

Table 4. Dynamics of the parameters characterizing the state of lipid component of fat-free and lecithin enriched soybean flour during storage in different types of packing

Duration of storage, Months	Parameter	Multi-layer (aluminum) packing, (M±m)	Mono-layer (transparent) packing (M±m)
0	[TBA-RS], nmole/mg of protein	0.330±0.042 (n=6)	
	[ROOH], mmole/g of LH	0.0053±0.0007 (n=3)	
	[TL]/ ADM, g/g	0.043±0.009 (n=3)	
1	[TBA-RS], nmole/mg of protein	0.296±0.074 (n=3)	0.324±0.010 (n=3)
	[ROOH], mmole/g of LH	0.0071±0.0008 (n=3)	0.0067±0.0006 (n=3)
	[TL]/ ADM, g/g	0.0346±0.0012 (n=3)	0.0423±0.0015 (n=3)
2	[TBA-RS], nmole/mg of protein	0.193±0.066 (n=3)	0.195±0.027 (n=3)
	[ROOH], mmole/g of LH	0.0036±0.0019 (n=3)	0.0044±0.0006 (n=3)
	[TL]/ ADM, g/g	0.044±0.005 (n=3)	0.0437±0.0025 (n=3)
3	[TBA-RS], nmole/mg of protein	0.610±0.103 (n=3)	0.816±0.150 (n=3)
	[ROOH], mmole/g of LH	0.0043±0.0013 (n=3)	0.0042±0.0002 (n=3)
	[TL]/ ADM, g/g	0.055±0.011 (n=3)	0.058±0.0015 (n=3)

*n – number of parallel measurements.

During storage of lecithin enriched soybean flour in the multi-layer (aluminum) packing, the increase of relative contents of PC by 1.3 times, the decrease of the CL+PA by 1.9 times as well as the tendency of the PI+PS fraction content decrease are observed. The more significant changes in the relative contents of the PL minor fractions are observed when storing the flour in mono-layer (transparent) packing. After three months of storage, the proportion of lysoforms in phospholipids is increased by 1.5 times and the relative contents of SM is increased by 2.85 times as well as the decrease of the proportion of PI+PS sum fraction is observed. Besides, the appearance of an insignificant amount of PG in the soybean flour PL composition in both types of packing after one month of storage should be pointed out.

The dynamics of the state of the generalized parameters of the fat-free soybean flour lipids composition during storage is presented in Figures 1 -5.

The data presented show that a typical decrease of the ratio between the sums of the more easily oxidizable to the more poorly oxidizable PL fractions (Figure 5), the increase of the PC/PE ratio in phospholipids (Figure 4) and insignificant variations of the sterol content in the composition of total lipids (Figure 2) during storage of fat-free lecithin enriched soybean flour can be observed. The type of packing practically not affecting the sterol content (Figure 2), the ∑EOPL / ∑POPL ratio (Figure 5) during three months of storage and the PC/PE ratio (Figure 4) during the first two months of storage. Three months later, the PC/PE ratio when storing the product in a multi-layer (aluminum) packing overtops the respective index in comparison to storing the product in a mono-layer (transparent) packing. Besides, storage in the multi-layer (aluminum) packing contributes to better keeping of phospholipids (Figure 1) in the total lipids whereas the [sterols] / [PL] ratio is more stable when storing the product in a mono-layer (transparent) packing (Figure 3).

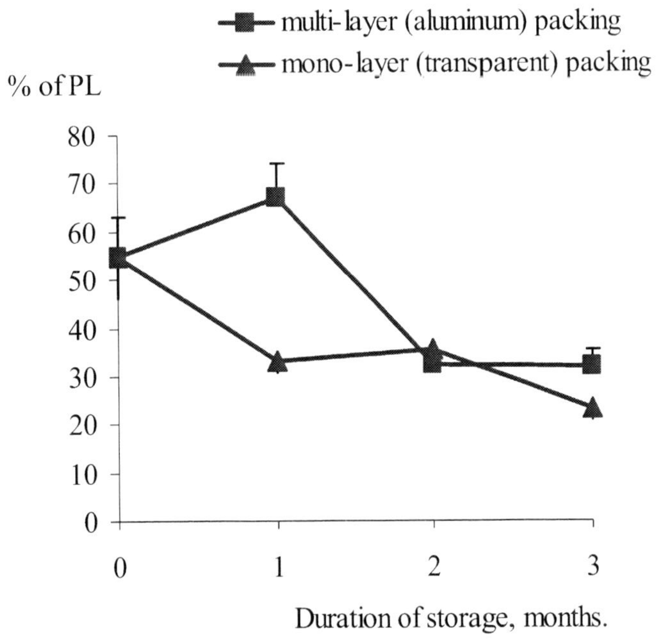

Figure 1. Effect of packing type on the content of phospholipids in the total lipids of fat-free and lecithin enriched soybean during storage.

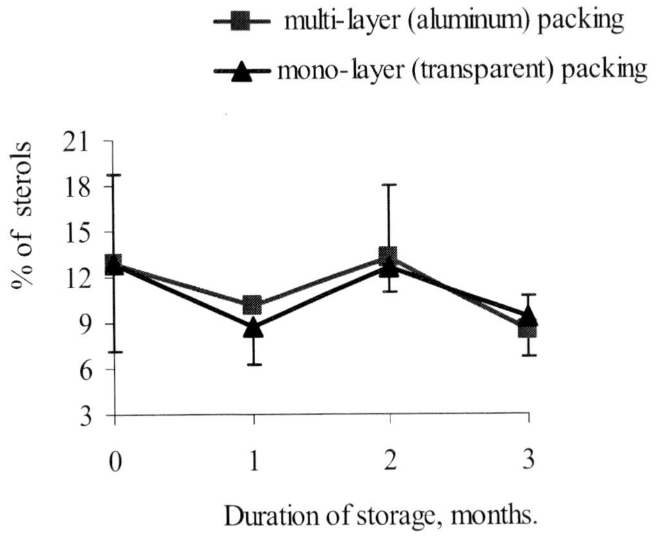

Figure 2. Effect of packing type on the content of sterols in the total lipids of fat free and lecithin enriched soybean during storage.

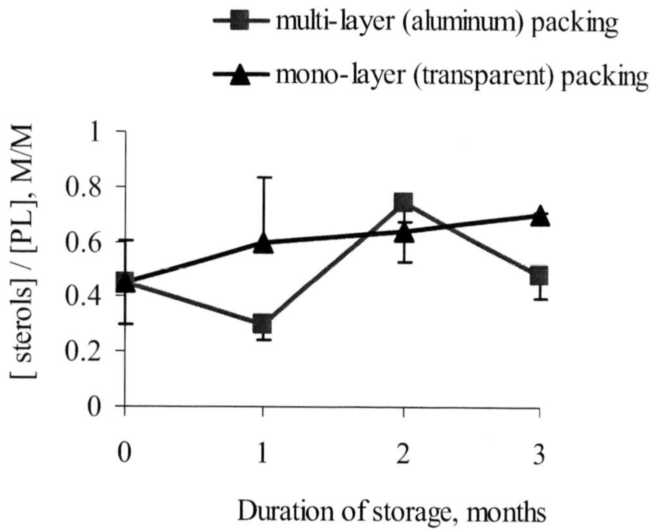

Figure 3. Effect of packing type on [sterols]/[phospholipids] ration in lipids of fat free and lecithin enriched soybean flour during storage.

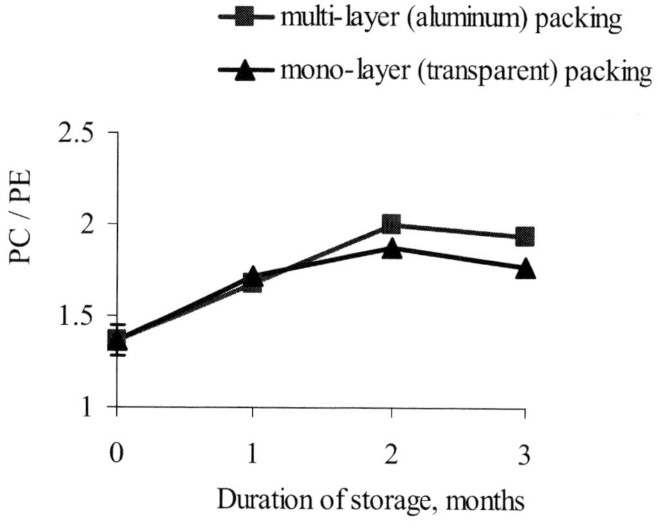

Figure 4. The ration of phosphatidyl choline / phosphatodyl ethanolamine in phospholipids content of fat free and lecithin enriched soybean flour during storage in different types of packing.

The total lipids contents in the storage process as a whole do not significantly differ as compared to their amount both in the original flour and in the process of storage in different types of packing (Table 4). However, after three months of storage in a mono-layer (transparent) packing, the amount of substances extracted by the chloroform/methanol mixture has a tendency of growing by approximately 35%. The amount of peroxides in lipids has a tendency of increasing after one month of storage and is slightly reduced within subsequent period irrespective of the type of packing (Table 4). Much more variable is the [TBA-RS] contents, which is slightly reduced during the first two months of storage and credibly increases by 1.85 times after three months of the experiment when the product is stored in a multi-layer (aluminum) packing and by 2.5 times when stored in mono-layer (transparent) packing (Table 4).

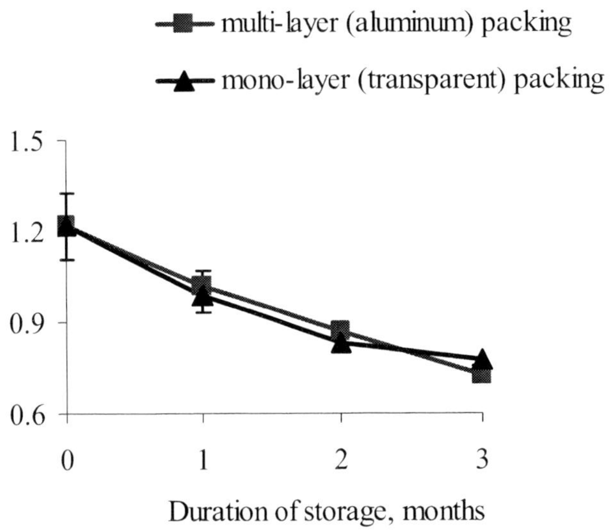

Figure 5. The sums of the more easily oxidizable and the more poorly oxidizable fractions in phospholipids (∑EOPL/∑POPL) of fat free lecithin enriched soybean flour during storage in different types of packing.

Thus, the process of storage of fat-free soybean flour with addition of lecithin SOPROPOLETS-8-TB-325 causes credible changes of a large variety of parameters of its lipid component that have been already discovered one month after the beginning of the experiment. This may be due to a high content of polyunsaturated fatty acids consisting to 48.3% in semi-fat free soy flour [10]. The necessity of stabilization of the of soybean flour lipid component requires further investigation for selecting efficient antioxidants allowing to increase its keeping period.

The results obtained demonstrated the absence of the essential differences of the parameters in the lipid component of fat-free soybean flour with addition of lecithin during it's storage both in the multi-layer (aluminum) and mono-layer (transparent) packing.

REFERENCES

[1] I.M. Emanuel, J.N. Lyaskovskaya *"Inhibition of Fat Oxidation"*. Moscow: Pishepromizdat, 1961, 355 pp. (in Russian).

[2] F.M. Rzhavskaya *"Lipids of fishes and marine mammals"*. Moscow: Pishepromizdat, 1976, 470 pp. (in Russian).

[3] I.N. Gulyaev, T.S. Zakhartchenko, D.I. Kuznetsov, S.J. Lazareva, T.F. Roenko, M.E. Konchalovskaya, L.M. Radtchenko *"Investigation of natural antioxidants of fat stabilizers for food concentrates"*: Improvement of Production of food concentrates. Moscow: CNIITEI, 1985. pp. 73-81. (in Russian).

[4] A.N. Dorohovitch, N.V. Alekseenko, R.N. Krimarenko, S.G. Prudnikov, G.N. Trembovetskaya// *Breadbaking of Russia*. 1998. No. I. pp. 23-24. (in Russian).
[5] L.N. Shishkina, M.A. Klimova, G.F. Dremutcheva, S.E. Traubenberg. // *Applied biochemistry and microbiology*. 2000. V. 36. No. 4. pp. 434 - 438.
[6] V.L. Kretovitch *"Biochemistry of grain and bread"*. Moscow: Nauka, 1991. 136 pp. (in Russian).
[7] V.A. Menshov, L.N. Shishkina, Z.N. Kishkovskii // *Applied biochemistry and microbiology*. 1993. V. 29. No. 6, pp. 675 - 683.
[8] A.B. Lisitcin, B.E. Gutnik, I.G. Anisimova, M. Smirnov (VNII (All-Russian Scientific Research Institute) of meat industry)) M. Ikatch, V.I. Malikava (VOBEKS-INTERSOYA Company). // *Foof industry* No. 4, 2002, pp.50-52. . (in Russian).
[9] M.L. Domoroshenkova // *Food industry,* 2001, No. 4, pp. 5-11. (in Russian).
[10] V.V. Kolpakova, I.V. Martynova, L.I. Arabova, L.V. Tchumikina // *Applied biochemistry and microbiology*. 2004. V. 40. No. 6. pp. 693 - 698.
[11] A.S. Buldakov. *Food additives. Reference book – St. Petersburg, «Ut»*, 1996. 240 pp. (in Russian).
[12] Certificate of correspondence No. ROSS YU. ПР78. H13631.
[13] T. Asakawa, S. Matsushita. *Lipids*. - 1980. - V. 15. - No. 3. pp. 137-140.
[14] R. Itzhaki, D.M. Gill. *Anal.Biochem*.- 1964.- V. 9. pp. 401 - 409.
[15] M. Kates. *The technique of lipidology*. Moscow: Mir, -1975. 322. pp. (Russian version).
[16] "*Biological membranes. A practice approach.*" / Eds J. B.C. Findlay, W.H. Evans – Moscow: Mir,-1990. 424 pp. (Russian version).
[17] W.M. Sperry, M. Weeb. *Biol.Chem*. - 1950. V. 187. - No. 1. pp. 97-106.
[18] L.N. Shishkina, E.V. Kushnireva, M.A. Smotrjaeva. *Radiat. biology. Radioecology*. 2001. - V. 41. No. 3. pp. 301-306. (in Russian).
[19] G.F. Lakin. *Biometry*. 3rd publication. Moscow: Higher school, 1990. 293 pp. (in Russian).

In: New Trends in Biochemical Physics Research
Editors: S. D. Varfolomeev et al., pp. 111-119
ISBN 1-60021-463-0

Chapter 12

THE ANALYSIS OF THE CONSTRUCTIONS OF THE MAGNETIC SEALS WITH HONEYCOMB ELEMENTS

F. S. Bayburtskiy[1*] and P. S. Kirey[2]
[1]N. M. Emanuel Institute of Biochemical Physics, Russian Academy of Sciences, Russian Federation, Moscow, 119991, Kosygina street, 4.
[2]Ukrainian State Maritime Technical University, P. O. Box 562, Nikolajev 54031, Ukraine

ABSTRACT

It is survey of the new technological and constructional solutions for improvement magnetic fluid seals for vacuum techniques. Survey includes information about working out and receiving of high steady magnetic fluids on the basis of the vacuum oils and other liquids with low pressure of saturated vaporous. High stability of the made magnetic fluids is provided with using of the natural and synthetically organic acids as stabilizers of the colloid system. Survey contains the analysis of the constructions of the four types of magnetic fluid seals: with sunk magnetic polar tip, open-frame, having honeycomb elements, and regulative. New technological and constructional solutions allow to decrease mass-overall characteristics and cost, and to raise reliability of the work of the magnetic fluid seals too. It makes them more attractive for using in industrial production and scientific researches.

Keywords: Magnetic fluid seals; honeycomb elements; sunk magnetic polar tip

THE REVIEW AND DISCUSSION

The combining of the magnetic fluids seals (MFS) with the other types of seals allows in a great number of cases to enlarge the diapason of the working parameters and branches of

* E-mail: Chembio@sky.chph.ras.ru, Bayburt@mail.ru

the application of the MFS [1 – 17]. The application of structural elements looking like honeycomb fillers in MFS is one of the perspective directions of the combining. In this case positive properties of honeycomb seals are partly used, these are:

- the high sealing capacity, because they can work with very minor gaps;
- reduced requests of the precision of the centering of the sealing relatively to the shaft;
- the absence of the risk of the mechanical injury of the shaft, because honeycomb filler can easily collapse at the emergency operating mode;
- the possibility of the realization of the honeycomb filler with various size of cells for the improvement of the hydrodynamic characteristics of the layer of the sealing medium in the running clearance;
- the possibility of the realization of walls of the cells of the honeycomb filler with the various thickness for the optimization of the detained pressure differential;
- the possibility of the realization of the composite (sectional) honeycomb filler.

It is not necessary to thread magnetic flux stages on the sunk magnetic polar tip when they look like the honeycomb fillers, because walls of the cells act like stages. Besides the increasing of the reliability is reached, because cellular structure of the magnetic fluid allocation is formed in the working clearance instead of the separate magnetic fluid «liquid in rings» of the sealing [1 – 3].

The bursting of this structure has a local disposition, because sealed medium, whish is under pressure, bursts the only cell. This leads to the faster localization of the place of bursting and to its self healing. The presence of the additional quantity of the magnetic fluid in the cells of the honeycomb fillers also raises reliability MFS and increases its resource characteristics in a great extent [4 – 5].

It is expedient to apply the alloys of permalloy type for making of the honeycomb sunk magnetic polar tip, because these alloys have a maximal value of the magnetic penetrability, what allows to decrease to minimum losses of the magnetic flux in sunk magnetic polar tip and decrease to minimum the necessary thickness of the walls of the honeycomb filler's cells. The optimal thickness of the walls of the honeycomb filler's cells is the diapason 0.1 – 1 mm, the lower limit of the diapason is determined by technological possibilities to making of the honeycomb filler and by the magnetic saturation of the material, the higher limit is defined by the decreasing of the magnetic induction in the working clearance and, therefore, by the decreasing of the detained pressure differential [6 – 8].

The installation of the non-magnetic honeycomb ring-shaped element in the between poles space of the MFS is necessary to placing the reserve quantity of the magnetic fluid inside the element. This gives the additional possibility to increasing of the reliability and resource of the MFS work. Usually the non-magnetic porous material, which is saturate with magnetic fluid, applies to this purpose. However the porous material, which fills the space between sunk magnetic polar tip, has the considerable specific surface which prevents the transference of the magnetic fluid to the working clearance in the case of necessity of the charging of working clearance by the magnetic fluid. This circumstance and the isotropy of the properties of the porous material lead to the appearance of the stagnant zones in its thickness. The magnetic fluid from these zones cannot get into the working clearance. The making of the additional tank for magnetic fluid looking like the honeycomb filler allows to remove this defect. Because the honeycomb capillary structure has less progressive specific surface than the porous material, the magnetic fluid can move from thickness of the honeycomb filler to the working clearance at the lesser values of the gradient of the magnetic field tensity, that is why the conditions for appearance of the stagnant zones in the thickness of the honeycomb filler are absent. The strong anisotropy of the properties of thy honeycomb filler in the axled and radial directions promotes to this [9 – 10].

The technology of making of the honeycomb fillers lets to produce sunk magnetic polar tip and non-magnetic ring-shaped element as the one detail, the middle part of which are non-magnetic and it is placed between sunk magnetic polar tip. It provides simplification of the construction of the seal and increase of the technologies of assemblage. For the decreasing of the magnetic flux in the place between shaft of cells the partial or complete filling of the cells by the non-magnet material is possible. Because the dispersion coefficient is proportionate to the value of the magnet penetrability, hence in this case the decreasing of the dispersion of the magnetic flux in 1.1 – 3 times takes place [11 – 12].

The three variants of the filling of the honeycomb elements may be. These are: 1 – cells are completely filled by the magnetic fluid, in this case sunk magnetic polar tip are the additional tank for the magnetic fluid, it raises the reliability of the MFS and increases in a great extent its resource characteristics; 2 – cells are completely filled with non-magnetic material, in this case dispersion magnetic fluxes decreased and accordingly the detained pressure differential increases and the turned to the working clearance surface of sunk magnetic polar tip is smooth, that decrease the own friction moment and increase the permissible velocity of the shaft rotation; 3 – the intermediate variant of the partial filling of the cells with magnetic fluid and nonmagnetic material [13 – 14].

The analysis of the hydrodynamic processes shows that the additional activation of the magnetic fluid happens in the dynamic working regime MFS at the performance of the constructive elements of the MFS as the honeycomb fillers having cellular structure, but this is connected with the increasing of the own moment of friction and with the dissipative losses, which decrease the permissible velocity, which decrease the permissible velocity of the rotation of the shaft. It is expedient complete filling of the cells of the honeycomb filler with non-magnetic material which has anti-friction properties for the decreasing of the dissipate losses. The smoothing of the inner surface and the decreasing of the friction coefficient of the pair «inner surface» – «magnetic fluid» to decrease the own friction moment is several times. But in this case the honeycomb filler ceases to perform the function of the additional tank for the magnetic fluid.

All known designs of magnetic fluid seals are shown in figures 1 – 10.

The application of the MFS with the honeycomb elements is perspective at the relatively slow velocities of the rotation of the shaft (to 20 m / s) and at the low-pressure differential (to 0,5 MPa), but at the increased requests of the reliability and working resource [15 – 17].

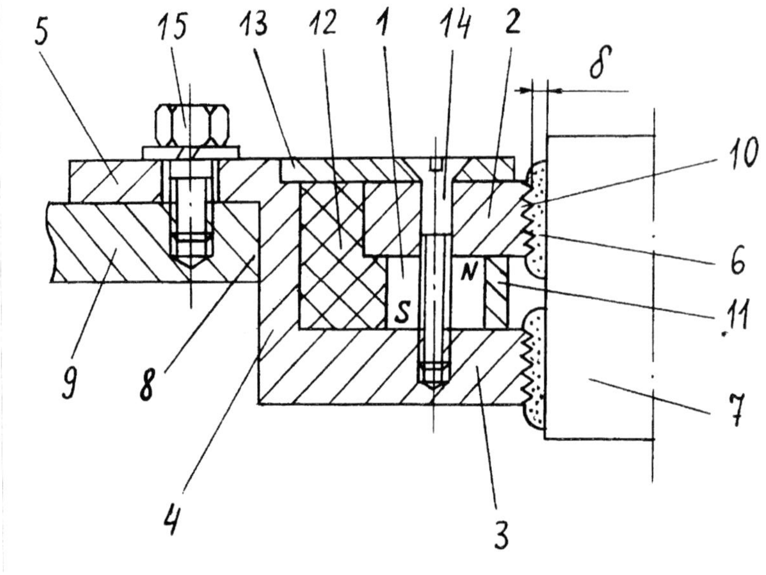

Figure 1. MFS with a sunk magnetic polar tip (variant № 1): δ - working backlash, 1 - permanent magnet, 2 - ring polar tip, 3 - internal ring ledge, 4 - casing, 5 - external connecting flange, 6 – magnetic fluid, 7 - shaft, 8 - ring landing element, 9 - the case of a drive, 10 - concentrators of a magnetic stream, 11 - not magnetic ring, 12 - not magnetic filler, 13 - not magnetic cover, 14 - not magnetic fixing element, 15 - fixing element.

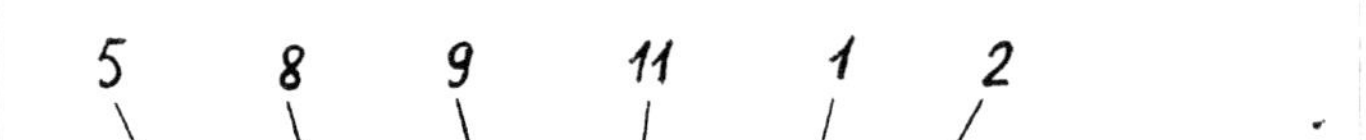

Figure 2. The MFS with a sunk polar tip (variant № 2): δ - working backlash, 1 - permanent magnet, 2 - ring polar tip, 3 - ring landing element, 4 - connecting flange, 5 - board, 6 – magnetic fluid, 7 - shaft, 8 - not magnetic cartridge, 9, 10 - through passage aperture, 11 - not magnetic fixing element, 12 - the case of a drive, 14 - not magnetic ring.

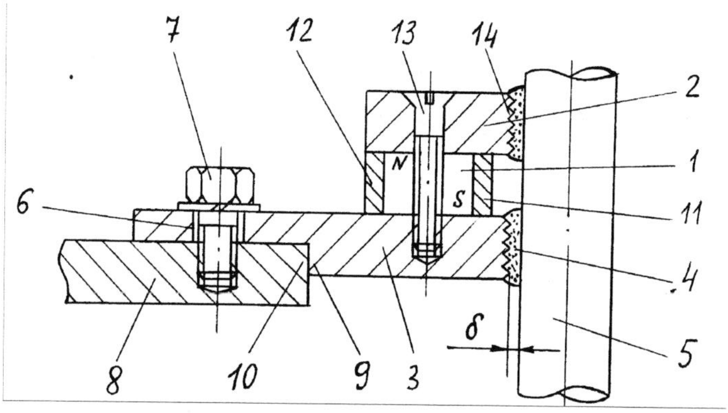

Figure 3. Open MFS with a polar magnetic tip as a connecting flange: δ - working backlash, 1 - permanent magnet, 2 - polar tip, 3 - polar tip as a connecting flange, 4 – magnetic fluid, 5 - shaft, 6 - through passage aperture, 7 - fixing element, 8 - the case of a drive, 9, 10 - landing element, 11 - internal not magnetic ring, 12 - external not magnetic ring, 13 - not magnetic fixing element, 14 - concentrators of a magnetic stream.

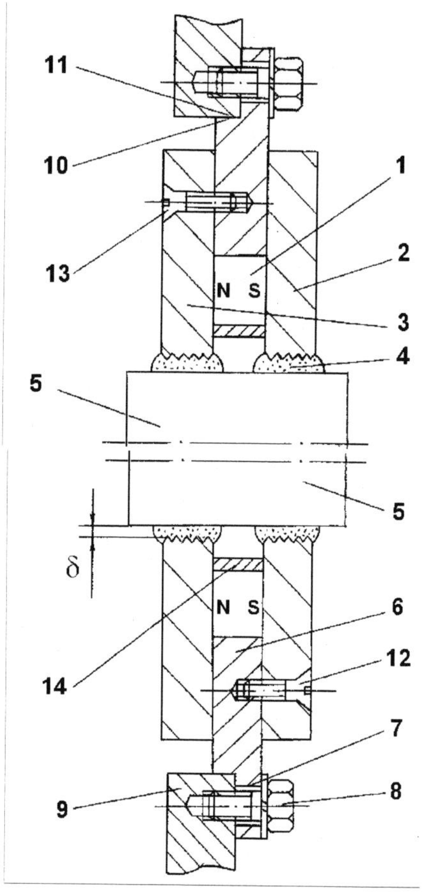

Figure 4. Open MFS with not magnetic cartridge as a connecting flange: δ - working backlash, 1 - permanent magnet, 2,3 - polar tip, 4 – magnetic fluids, 5 - shaft, 6 - the ring not magnetic cartridge, 7 - through passage aperture, 8, 12, 13 - fixing element, 9 - the case of a drive, 10, 11 - landing element, 14 - not magnetic ring.

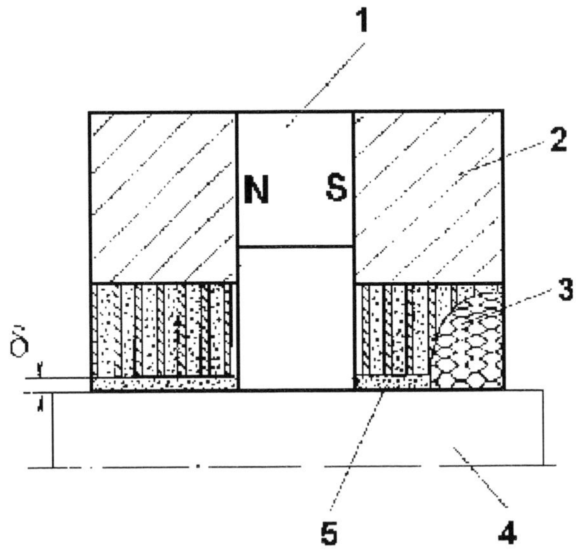

Figure 5. The scheme of application of cellular elements in MFS with axial magnetized inductor: δ - working backlash, 1 - constant magnet, 2 - magnetic permeable ring, 3 - honeycomb ring element, 4 - shaft, 5 - magnetic fluid.

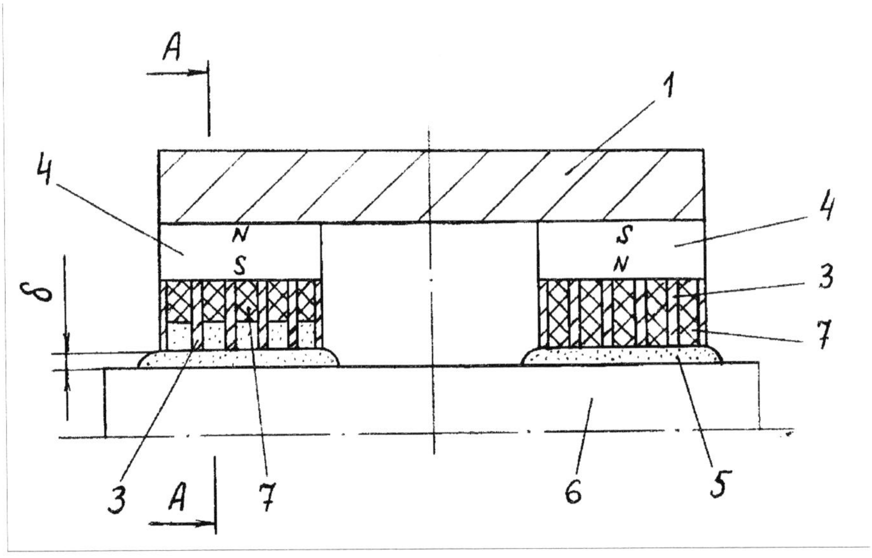

Figure 6. The circuit of application of cellular elements in MFS with radial magnetized inductor: δ - working backlash, 1 - closing magnetic conductor, 3 - honeycomb magnetic permeable element, 4 - permanent magnet, 5 - magnetic fluid, 6 - shaft, 7 - not magnetic filler.

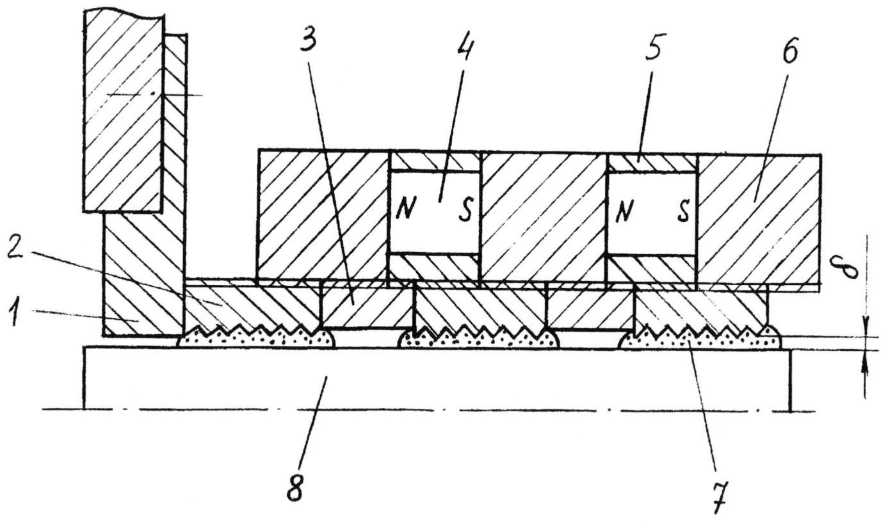

Figure 7. The MFS with demountable a closing magnetic conductor and permanent magnets: δ - working backlash, 1 - connecting flange, 2 - polar tip, 3 - not magnetic cartridge, 4 - demountable permanent magnet, 5 - demountable closing magnetic conductor, 6 - magnetic fluid, 7 - shaft, 8 - refueling aperture, 9 - stopper a refueling aperture.

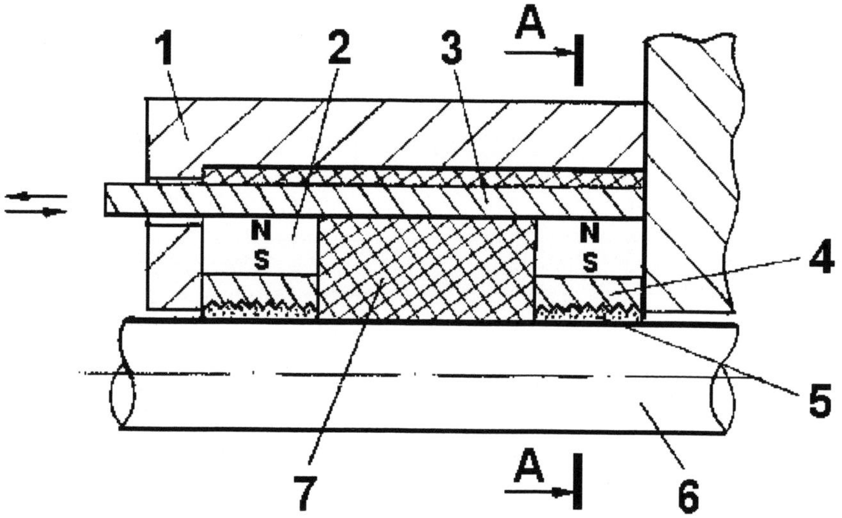

Figure 8. The MFS with a mobile closing magnetic conductor: 1 - the case, 2 - permanent magnet, 3 - closing magnetic conductor, 4 - polar tip, 5 – magnetic fluid, 6 - shaft, 7 - not magnetic filler.

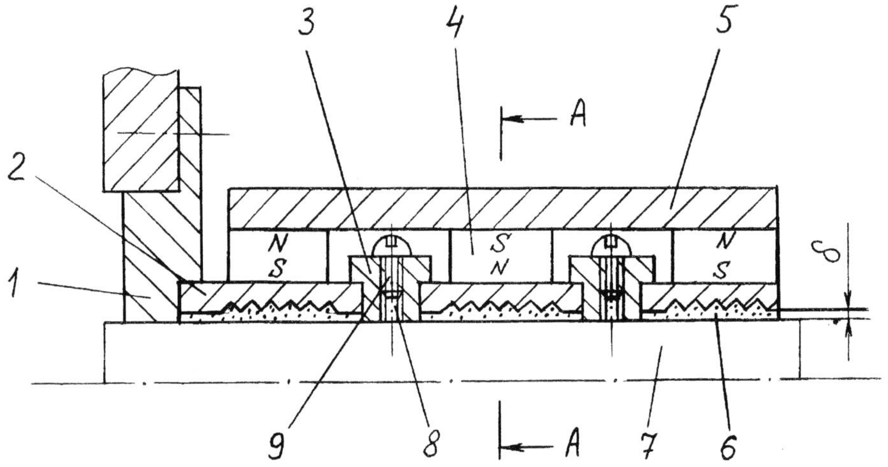

Figure 9. The MFS with mobile magnetic system: δ - working backlash, 1 - connecting flange, 2 - motionless part of a polar tip, 3 - not magnetic cartridge, 4 - permanent magnet, 5 - not magnetic holder, 6 - mobile part of a polar tip, 7 – magnetic fluid, 8 – shaft.

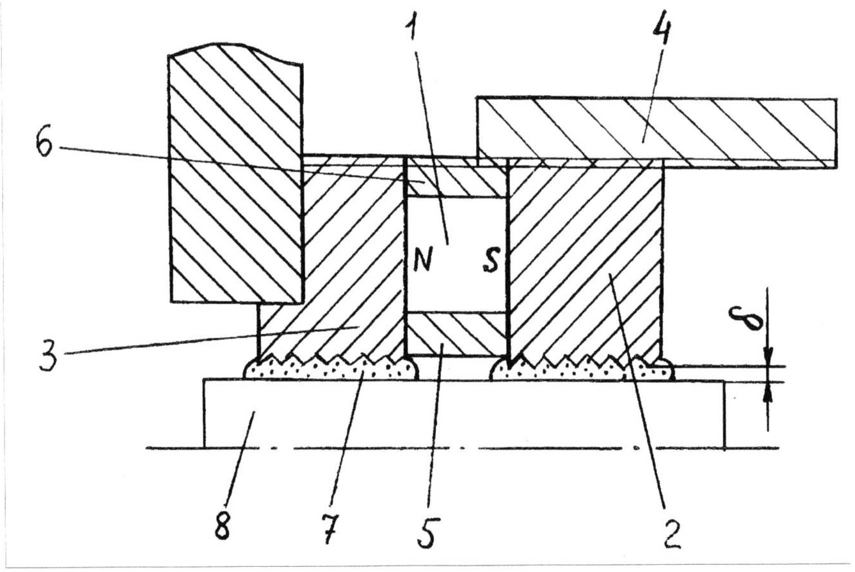

Figure 10. The MFS with a mobile shunting element: δ - working backlash, 1 - permanent magnet, 2, 3 - polar tip, 4 - shunting element, 5 - internal not magnetic ring, 6 - external not magnetic ring, 7 - magnetic fluid, 8 – shaft.

Conclusion

The magnetic fluid seals a lot of advantages in comparison with traditional systems of hermetic sealing and are widely applied in industrially advanced countries of the world. In the present work ways of improvement vacuum magnetic fluids seals are analyzed with the purpose of their adaptation to existing conditions of industrial production in Russia, Ukraine and Belarus. Industrial technologies inexpensive highly stability magnetic fluids are developed on the basis of vacuum oils and other fluids with low pressure sated pairs. The analysis of designs of four types of magnetic fluid seals is lead: with a sunk polar tip, with cellular elements and adjustable. New technological and constructive decisions allow to reduce mass-dimensional characteristics and cost, and also to raise reliability of work of magnetic fluids seals, that does their more attractive to use in industrial production and scientific researches.

REFERENCES

[1] J. L Neuringer, R.E. Rosenszweig. // *Phys. Fluids* 7 (1964) 966.

[2] S. S. Papell. Low viscosity magnetic fluid obtained by the colloidal suspension of magnetic particles / Patent USA № 3215572, 1965.

[3] R. E. Rosensweig, G. Miskolczy, F. D. Ezekiel. // *Machine Design.* 40 (8) (1968) 145.

[4] D. V. Orlov. Magnetic vacuum seals. / Patent of USSR № 420836. Bull. Isobret. № 11, 1974.

[5] P. L. Windle, J. Popplewell, S. W. Charles. // *IEEE Trans. Magn.* Mag -11 (1975) 1367.

[6] M. Goldowsky. // *IEEE Trans. Magn.* Mag-16 (1980) 382.

[7] R. Massart. // *IEEE Trans. Magn.* Mag-175 (1981) 1247.

[8] Nakatani, T. Furubayashi, T. Takahashi. // *J. Magn. Magn. Mater.* 65 (1987) 261.

[9] G.T.Y. Wan, H.A. Spikes. // *STLE Trans.* 31 (1988) 12 – 21.

[10] S. Lingard, W.A. Bullough, W.M. Shek. // *J. Phys.* D 2 (1989) 1639 – 1645.

[11] Nakatani, M. Hijikata, K. Ozawa. // *J. Magn. Magn. Mater.* 122 (1993) 10.

[12] J. Anon. User manual of TE53SLIM multi-purpose friction and wear tester. / PLINT and Partners Ltd., Version 3.8, 1995.

[13] T. Atarashi, Y.S. Kim, T. Fujita, K. Nakatsuka. // *J. Magn. Magn. Mater.* 201 (1999) 7 – 10.

[14] T. Atarashi, Y. S. Kim, T. Fujita. // *J. Magn. Magn. Mater.* 201 (1999) 361 – 363.

[15] J. D. Carlson. Magneto rheological fluid actuators, in: H Janocha (Ed.), *Adaptronics and Smart Structures, Springer*, Berlin, 1999, Chapter 6.6, P. 180–195, ISBN 3-540.61484-(26).

[16] P. L. Wong, W. A. Bullough, C. Feng, S. Lingard. // *Wear.* 247 (2001) 33 – 40.

[17] H. Nishiyama, T. Oyama, T. Fujita. // *International Journal of Modern Physics* B, 15 (6 – 7) (2001) 829 – 836.

INDEX

A

B

C

D

E

F

G

H

I

J

K

L

P

Q

R

S

T

U

V

W

Y